International Research in Science and Soccer

International Research in Science and Soccer showcases the very latest research into the world's most widely played sport. With contributions from world-leading researchers and practitioners working at every level of the game, from grass roots to elite level, the book covers every key aspect of preparation and performance, including:

* Contemporary issues in soccer coaching
* Psychological preparation and development of players
* Physical preparation and development of players
* Nutrition and recovery
* Talent identification and development
* Strength and conditioning in soccer
* Injury prevention and rehabilitation
* Soccer academies.

Sports scientists, trainers, coaches, physiotherapists, medical doctors, psychologists, educational officers and professionals working in soccer will find this in-depth, comprehensive volume an essential and up-to-date resource.

The papers contained within this volume were first presented at the First World Congress on Science and Soccer, held in May 2008 in Liverpool, UK. The meeting was held under the auspices of the World Commission of Science and Sports.

Barry Drust is a Reader in Applied Exercise Physiology at the Research Institute for Sport and Exercise Sciences, Liverpool John Moore's University. His main research interests are the physiology of intermittent exercise and temperature regulation. He has published numerous books and journal articles on the physiology of soccer players.

Thomas Reilly was Director of the Research Institute for Sport and Exercise Sciences at Liverpool John Moore's University from 1996 to 2009. He is the Chair of the International Steering Group for Science and Football, and for 13 years was Chair of the Exercise Physiology Steering Group of the British Olympic Association. He was awarded a Doctor of Science degree for his research work in 1999 and has worked in sports science support roles with professional football teams and Olympic Games athletes.

A. Mark Williams is Professor of Motor Behaviour at the Research Institute for Sport and Exercise Sciences, Liverpool John Moore's University. He has published widely in areas related to motor control and learning, sport psychology, performance analysis and talent identification.

International Research in Science and Soccer

The Proceedings of the First World Conference on Science and Soccer

Edited by
Barry Drust, Thomas Reilly and A. Mark Williams

Routledge
Taylor & Francis Group

LONDON AND NEW YORK

First published 2010 by Routledge

2 Park Square, Milton Park, Abingdon, Oxfordshire OX14 4RN

Simultaneously published in the USA and Canada
by Routledge

711 Third Avenue, New York, NY 10017

Routledge is an imprint of the Taylor & Francis Group, an informa business

First issued in paperback 2011

Typeset in Times New Roman for the authors by Zoe Miveld

British Library Cataloguing in Publication Data
A catalogue record for this book is available from the British Library

Library of Congress Cataloging in Publication Data
International research in science and soccer / edited by Barry Drust,
Thomas Reilly and A. Mark Williams.
    p. cm.
    Includes index.
    1. Soccer. 2. Sports sciences. I. Drust, Barry. II. Reilly, Thomas, 1941-III.
Williams, A. M. (A. Mark), 1965-
    GV943.I54 2010
    796.334—dc22

2009017535

ISBN 13: 978-0-415-49794-7 (hbk)
ISBN 13: 978-0-203-87750-0 (ebk)
ISBN 13: 978-0-415-67733-2 (pbk)

Contents

Preface

This book represents the Proceedings of the First World Conference on Science and Soccer held at Liverpool, May 15-16, 2008. The event was held under the auspices of the World Commission for Science and Sports and specifically its International Steering Group on Science and Football. It was focused on the game of association football (soccer) and so differed from previous conferences held under the aegis of the International Steering Group.

The first World Congress on Science and Football held at Liverpool in 1987 represented a watershed in the application of science to the football codes. It was the first ever occasion that scientists mingled with formal representatives of the international and national codes of football; these included soccer, rugby union, rugby league, American football, Australian Rules football and Gaelic football. The Proceedings of this Congress included 88 contributions and ran to 651 pages. The book is now a collector's item that is still much sought after; it was recently advertised on the website of Amazon for £420. These congresses are held over 5 days every 4 years; further meetings were held at Eindhoven (Netherlands, in 1991), in Cardiff (Wales) in 1995, in Sydney (Australia) in 1999, in Lisbon (Portugal) in 2003 and in Antalya (Turkey, 2007). The Seventh World Congress on Science and Football is scheduled for Nagoya (Japan) in 2011.

The aims from the outset for the 'science and football' congresses were to bridge the gap between theory and practice in the various football codes and transfer knowledge across these different forms of football. As scientific findings have been increasingly incorporated into football practice, the majority of professional soccer clubs began to employ sports science personnel to provide counselling services and support for their work. This development caused an increased demand for up-to-date scientific knowledge to provide solutions to problems in the applied context of soccer practice. As circumstances and culture within the clubs changed, there was an accompanying growth in research in applied settings. These developments were the drivers for this congress targeted at soccer specifically and based on current concerns within the game and its overall environment.

The First World Conference on Science and Soccer was a two-day event that allowed practitioners working within professional soccer to come together to share their ideas and applied knowledge. Three Workshops, each repeated once, six invited keynote lectures at three separate Plenary Sessions, and twelve invited presentations to Symposia on contemporary hot topics complemented the formal scientific communications. The 52 oral and poster communications enabled delegates from academic and club settings to report their most recent findings relevant to practice and form the content of this book. In order to maintain a programme of continuing professional development for those studying, working in, and researching into soccer, a conference every two years is intended, intercalated with the less frequent and more comprehensive meetings concerned with all the football codes. The Second World Conference on Science and Soccer is scheduled for the Republic of South Africa, immediately preceding the FIFA World Cup.

The conference in Liverpool in 2008 coincided with the city being the European Capital of Culture in 2008. Altogether 17 countries outside the United Kingdom were represented on the programme. Throughout the year tourists visited the city in unprecedented numbers to attend the various cultural events that marked this unique civic award. It was a source of pride to the organisers at Liverpool John Moores University that 'science and soccer' was linked to the programme of scientific events for the year. Its inclusion serves to emphasise the importance of soccer in the cultural life of people in the North West of England and the opportunities for research that the game provides on an international basis.

Thomas Reilly

Chair, International Steering Group on Science and Football

Introduction

This book represents the Proceedings of the First World Conference on Science and Soccer. This inaugural event was held over May 15-16, 2008 under the aegis of the International Steering Group on Science and Soccer, and is an event that is planned to occur every two years. The Conference attracted keynote and invited lectures, oral communications and poster presentations. These events were complemented by workshops and symposia on selected topics. It is the oral and poster material that forms the content of this book.

For inclusion in the programme, Abstract submissions were vetted by the scientific committee established for the meeting. Invitation for a full submission was extended to all presenters in the scientific sessions. These manuscripts were distributed among the editors and each manuscript was peer-reviewed by two referees. The content of these Proceedings is formed by those that passed through this quality control process successfully.

The content is divided into six Parts, roughly equivalent to the way the submissions were distributed to editors. These different sections are not mutually exclusive. For example the contributions in paediatric science might have been distributed elsewhere according to scientific discipline. It was an editorial decision to keep the content of research on young players together in one multidisciplinary section as far as possible. Similarly studies of body composition (including body composition of youth players) are alongside those in nutrition. Analysis of monitoring equipment is included with biomechanics where there is particular emphasis on research methods.

The content reflects the range of research activities currently engaging the science support personnel in soccer clubs and scientists concerned with the application of their work in a practical setting. It should be of interest to all students of science and football, sports science and related studies. It is immediately relevant to research workers in these areas. It is primarily of interest to practitioners and to trainers and coaches keen to increase their soccer-specific knowledge base.

We are grateful for the co-sponsorship of this conference provided by Jon Goodman, Director of Think Fitness (in association with Red Bull) and the associated sponsorship of EXF Fitness Equipment. The support of ten major exhibitors at the Conference was also appreciated in contributing to the professional parts of the overall programme. A special note of thanks goes to the staff members from the University's Conference and Events office who helped in organising the event. For the completion of this work and leaving a legacy of the event, we are indebted to the authors who responded positively to our queries and comments. We acknowledge the support of the various referees. We regret that many submissions could not be included due to space and other constraints. A special note of gratitude must be extended to the staff at the office of the Research

Institute for Sport and Exercise Sciences at Liverpool John Moores University and in particular to Zoe Miveld who helped with the technical aspects of translating the varieties of electronic material into one coherent text.

Editors: Barry Drust, Thomas Reilly and A. Mark Williams
Research Institute for Sport and Exercise Sciences,
Liverpool John Moores University,
March 2009

Part I
Paediatric Science

Somatic and maturity status of youth soccer players 11-12 years: Variation by field position

A. J. Figueiredo[1], M. J. Coelho e Silva[1] and R. M. Malina[2]

[1]Faculty of Sport Science and Physical Education, University of Coimbra, Coimbra, Portugal
[2]Department of Kinesiology and Health Education, University of Texas, USA

1. INTRODUCTION

Advanced maturation within the same year of birth, during the pre-pubertal and pubertal period, is associated with advantages in body size, percentage of fat-free mass and several functional manifestations such as aerobic performance, strength and velocity (Malina *et al.*, 2004a). Therefore, in sports where the above characteristics are factors that give an advantage in competition, the athletes maturationally more advanced within a chronologically equivalent group are favoured when compared with less developed peers. For example, Helsen *et al.* (2005) found an effect of the birth month in players selected for advanced training with a bias favouring the older ones (32.6% and 16.0% in the first and last trimesters respectively), in a male sample of Under12 and Under14 soccer players. Furthermore, for the senior English national soccer team, Richardson and Stratton (1999) reported that 50.0% of the participants in the final stage of the 1998 World Cup had been born in the first four official months of competition for youth soccer (between September and December). When the authors analysed the data based on the position of the players in the field, they reported that this discriminating effect was higher for goalkeepers, defenders and forwards than for midfield players.

In a study of 13 to 14-year-old sub-elite soccer players from the north of Portugal, Malina *et al.* (2004b) reported that midfield players seem to be the lightest and the shortest of all the players, the forwards being heavier and taller. Aziz *et al.* (2005) and Brocherie *et al.* (2005) observed the same trend in samples of adult soccer players. Gil *et al.* (2007) established the anthropometric and physiological profiles of young nonelite soccer players, between 14 and 21 years old, according to their playing position, the goalkeepers being the tallest and heaviest group.

The present study considers somatic and maturity characteristics of youth soccer players aged 11-12 years by field position. Consideration of the anthropometric characteristics in youth soccer, in general, and the playing position, in particular, and their association with maturational evaluation provide a contribution to the literature in this area.

2. METHODS

The sample of the present study comprised 87 male soccer players belonging to the 11 to 12 years old competitive age-group (Table 1). The organization of youth soccer in Portugal uses two-year age groups and all players were born in 1991 or 1992. Players were from 5 teams from the midlands of Portugal that compete at a regional level in a 9-month competitive season (September-May) through the Portuguese Soccer Federation. Players participated in three training sessions per week (each about 90 minutes) and one game per week, usually on Saturday. The study was approved by the Scientific Committee of the University of Coimbra and by each club. The players and their parents provided informed consent. Subjects were informed that participation was voluntary and that they could withdraw from the study at any time.

Height (m), body mass (kg), proportions (bicristal/biacromial ratio - %), adiposity (sum of triceps, subscapular, suprailiac and medial calf skinfolds - mm), skeletal age (SA) (years) evaluated by the FELS method (Roche *et al.*, 1988), chronological age (CA) (years) and the difference between SA and CA (SA minus CA) (years) were used to compare goalkeepers, defenders, midfielders and forwards. In the collection of data, the protocol described in Lohman *et al.* (1988) was followed. Data were converted to Z-scores and mean Z-scores were compared by position using descriptive statistics.

All data were collected within a two-week period under standard conditions in an indoor facility at the University of Coimbra. The physical measurements were obtained by an experienced researcher in anthropometry.

Table 1. Age (range, mean, standard deviation) and number of players assessed per field position.

		Chronological age			
	n	*Minimum*	*Maximum*	*Mean*	*Std. deviation*
Goalkeepers	8	11.17	12.45	11.71	0.47
Defenders	32	10.99	12.84	11.68	0.47
Midfielders	28	10.98	12.82	11.87	0.54
Forwards	19	11.04	12.94	11.92	0.61
Total	87	10.98	12.94	11.92	0.61

3. RESULTS and DISCUSSION

Descriptive statistics by position and results of ANOVA are presented in Table 2. In addition, players were profiled by position using Z-scores (Figure 1). The position related variation was significant for two variables: discrepancy between skeletal age and chronological age (F=2.74, P<0.05, η^2=9%) and height (F=3.67,

P<0.02, η^2=12%). Goalkeepers were larger in body size while forwards were taller rather than heavy but both were advanced in skeletal maturation (SA-CA). On the other hand, the midfielders tended to be shorter and lighter and slightly later in skeletal maturation. Besides goalkeepers being younger, they were almost a full standard deviation above their peers, in general, in skeletal maturation and height. The defenders were the fatter and the youngest field-position group. Although speculative, a possible explanation for these results may be differences between central and lateral defenders, the former group being fatter and more mature and the latter group younger and less advanced in maturity.

Table 2. Z-scores of the sample, by field position, for the variables of the study.

	Goalkeepers (n=8)	Defenders (n=32)	Midfielders (n=28)	Forwards (n=19)	F (p)	η^2
Chronological age (CA), years	11.7 (0.5)	11.7 (0.5)	11.9 (0.5)	11.9 (0.6)	1.10 (0.35)	0.04
Skeletal age (SA), years	13.0 (1.1)	11.8 (1.4)	11.6 (1.4)	12.3 (1.5)	2.58 (0.06)	0.09
SA-CA difference, years	1.3 (1.2)	0.1 (1.4)	-0.3 (1.4)	0.4 (1.5)	2.74 (0.05)	0.09
Body mass, kg	41.1 (5.9)	38.7 (7.1)	36.3 (5.0)	38.4 (6.0)	1.53 (0.21)	0.05
Height, metres	1.498 (0.051)	1.447 (0.075)	1.420 (0.054)	1.463 (0.064)	3.67 (0.02)	0.12
Bicristal/biacromial ratio, %	69.6 (2.2)	71.6 (4.1)	73.6 (4.6)	72.3 (4.0)	2.31 (0.08)	0.08
Adiposity, mm	29.6 (9.8)	37.2 (17.4)	29.2 (13.1)	30.1 (10.8)	1.97 (0.12)	0.07

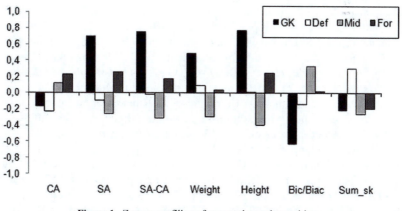

Figure 1. Z-score profiling of soccer players by position.

4. CONCLUSION

In this study differences in somatic characteristics and maturation by field position were observed. These results confirm partially the observations by Malina *et al.* (2004b) and Gil *et al.* (2007). The former authors also found the lowest values for height and weight in the midfield group while Gil *et al.* (2007) reported that the goalkeepers were the heaviest and tallest of all the players.

Our results suggest that larger, maturationally advanced players gravitate towards certain positions. Youth coaches tend to use larger, more mature players in positions where larger size is deemed an advantage. So, considering soccer as a sport where physical contact is important, coaches seem to use, in zones of the field near to the goal (and excluding the goalkeepers), the tallest and the heaviest players (defenders and forwards).

This trend for selection of position based on body size at these ages may bias a full development of young players once it is not possible for them to experiment in different positions and, concomitantly, in different tactical behaviours.

References

Aziz, A.R., Tan, F., Yeo, A. and Teh, K.C., 2005, Physiological attributes of professional players in the Singapore soccer league. In *Science and Football V*, edited by Reilly, T., Cabri, J. and Araújo, D. (London: Routledge), pp. 139-143.

Brocherie, F., Morikawa, T., Hayakawa, N. and Yasumatsu, M., 2005, Pre-season anaerobic performance of elite Japanese soccer players. In *Science and Football V*, edited by Reilly, T., Cabri, J. and Araújo, D. (London: Routledge), pp. 144-154.

Gil, S.M., Gil, J., Ruiz, F., Irazusta, A. and Irazusta, J., 2007, Physiological and anthropometric characteristics of young soccer players according to their playing positions: relevance for the selection process. *Journal of Strength and Conditioning Research*, **21**, pp. 438-445.

Helsen, W.F., Van Winckel, J. and Williams, A.M., 2005, The relative age effect in youth soccer across Europe. *Journal of Sports Sciences*, **23**, pp. 629-636.

Lohman, T.G., Roche, A.F. and Martorell, R., 1988, *Anthropometric Standardization Reference Manual*. (Champaign, IL: Human Kinetics).

Malina, R.M., Bouchard, C. and Bar-Or, O., 2004a, *Growth, Maturation, and Physical Activity*, 2nd ed. (Champaign, Illinois: Human Kinetics).

Malina, R.M., Eisenmann, J.C., Cumming, S.P., Ribeiro, B. and Aroso, J., 2004b, Maturity-associated variation in the growth and functional capacities of youth football (soccer) players 13-15 years. *European Journal of Applied Physiology*, **91**, pp. 555-562.

Richardson, D.J. and Stratton, G., 1999, Preliminary investigation of the seasonal birth distribution of England World Cup campaign players (1982-98). *Journal of Sports Sciences*, **17**, pp. 821-822.

Roche, A., Chumlea, W. and Thissen, D., 1988, *Assessing the Skeletal Maturity of the Hand Wrist – FELS Method,* (Springfield, Illinois: CC Thomas).

CHAPTER TWO

Physical test performance of elite Portuguese junior soccer players according to positional roles

Á. Ramos, P. Vale, B. Salgado, P. Correia, E. Oliveira, A. Seabra,
A. Rebelo and J. Brito

Faculty of Sports, University of Porto, Porto, Portugal

1. INTRODUCTION

Soccer is one of the most popular sports worldwide and a lot of research has been conducted on various aspects of this sport. Most of the relevant studies are based on elite or male participants (Mohr et al., 2005) and there is little research about youth soccer players. With the increasing emphasis on talent development and the minimization of potential injuries among adolescent athletes, there is a consensus between coaches and sport scientists that a comprehensive, sport-specific investigation would aid in more clearly defining present levels of physiological fitness and sports skills specific to youth soccer (Vanderford et al., 2004).

Players must possess moderate to high aerobic and anaerobic power, have good agility, joint flexibility and muscular development, and be capable of generating high torques during fast movements (Reilly et al., 2000). As soccer is a team sport, an efficient organization of the team is required for the optimal development of players' abilities, the control of opponents, and the successful resolution of a match. Players are placed in certain positions to fulfil specific tasks.

Significant differences in age, stature, body fat and body mass index have been recently identified between elite players of different playing positions suggesting that players with a particular size and shape may be suitable for the demands of the various playing positions (Bloomfield et al., 2005). Positional role appears to have an influence on total energy expenditure in a match, suggesting that different physical, physiological and bioenergetic responses are experienced by players of different positions (Reilly and Thomas, 1976; Di Salvo and Pigozzi, 1998; Stroyer et al., 2004; Pyne et al., 2006; Di Salvo et al., 2007). For example, greatest overall distances appear to be covered by midfield players who act as links between defence and attack (Reilly and Thomas, 1976; Tumilty, 1993; Wisloff et al., 1998; Rienzi et al., 2000; Di Salvo et al., 2007). The best performances in the velocity, agility and power tests were observed in the group of forwards (Tumilty, 1993; Wisloff et al., 1998; Gil et al., 2007).

Therefore, the purpose of the present study was to describe and compare performance in physical tests of youth elite soccer players according to positional roles.

2. METHODS

Eighty-three youth elite soccer players from Portuguese Under-19 first division, practising 4 or 5 times per week (6.92 ± 0.74 hours of training per week) participated in this study. Mean (standard deviation) for age (years), height (m) and body mass (kg) were respectively 18.2 (0.5); 1.760 (0.07); 72.0 (7.2). In accordance with time-motion analysis data of previous studies (Bangsbo *et al.*, 1991; Di Salvo and Pigozzi, 1998) four groups of positional roles were considered, namely full-back (n=13), midfielder (n=37), central defender (n=13) and forward (n=20). Goalkeepers were excluded.

Table 1. Mean values and standard deviations (SD) of age, height (m) and body mass (kg) according to positional roles.

	Full-back	*Midfielder*	*Central defender*	*Forward*
Age	18.2 ± 0.5	18.2 ± 0.4	18.2 ± 0.5	18.2 ± 0.6
Height	1.74 ± 0.06	1.74 ± 0.07	1.83 ± 0.03	1.74 ± 0.06
Body mass	70.0 ± 7.4	71.2 ± 6.7	78.0 ± 6.6	71.1 ± 7.0

The study was performed during Christmas and New Year holidays to guarantee players were free from school affairs. After a warm-up, all the players performed six physical tests on two occasions interspaced by one week. On day one, the players were evaluated in sprint performance (5 and 30 metres) and agility (T-test). On day two, vertical jump (Squat Jump and Counter-Movement Jump - CMJ) and intermittent exercise performance (Yo-Yo Intermittent Endurance Test - level 2; Yo-Yo IE2) were measured. Photoelectric cells (Speed Trap II – Browser Timing Systems) were used in sprint and agility tests. For vertical jump evaluations an Ergojump device (Digitime 1000, Digitest Finland) was used. The reliability of the physical performance tests was assessed using intraclass correlation coefficients (ICC). The ICC values were 0.98 (Yo-Yo IE2), 0.97 (5 and 30 metres sprint, Squat Jump and CMJ) and 0.95 (agility). The coefficients indicate high reliability.

Descriptive statistics were calculated by each positional role. Differences across all positional roles for each physical performance test were tested with a factorial ANOVA. The Scheffe multiple comparisons test was used to check for specific differences by positional role. Statistical significance was set at 0.05. SPSS 15.0 was used in all analyses.

3. RESULTS

Table 2 shows the results of the analysis of variance model used to compare the mean values of the various physical tests according to positional roles. The youth soccer players did not show any significant differences in performance on the physical tests between the different playing positions, except for the Squat Jump. The central defenders and forwards showed higher values in Squat Jump and CMJ than full-backs and midfielders. The full-backs and midfielders performed best in the sprint and Yo-Yo IE2 tests but differences were non-significant. Agility mean values were reasonably similar in all positional roles.

Table 2. Mean values, standard deviations (SD), F-test and P values for each physical performance test according to positional roles.

Physical tests	Full-back Mean±SD	Midfielder Mean±SD	Central defender Mean±SD	Forward Mean±SD	F	P
5 m sprint (s)	1.03 ± 0.06	1.05 ± 0.06	1.06 ± 0.06	1.06 ± 0.05	0.543	0.654
30 m sprint (s)	4.25 ± 0.18	4.30 ± 0.14	4.29 ± 0.08	4.26 ± 0.11	0.572	0.635
Agility (s)	8.90 ± 0.24	8.88 ± 0.23	8.86 ± 0.26	8.81 ± 0.24	0.420	0.739
Squat Jump (cm)	34.4 ± 5.6	36.1 ± 3.8	39.1 ± 2.6	38.3 ± 5.3	2.802	0.05
CMJ (cm)	36.7 ± 5.0	37.8 ± 4.0	40.6 ± 5.0	40.7 ± 5.1	2.486	0.069
Yo-Yo IE2 (m)	1420 ± 522	1464 ± 391	1353 ± 330	1330 ± 425	0.539	0.657

4. DISCUSSION

Soccer performance represents a composite output of elite physical characteristics that, in turn, depend upon a variety of anthropometric and physiological properties, as well as on training health status of the individual athlete (Macarthur and North, 2005). The problem is especially complex in soccer where performance itself is multifactorial (Reilly et al., 2000).

In the present study although there were no significant differences between positional roles except in Squat Jump and a trend in CMJ, central defenders and forwards showed a higher performance in these physical tests than full-backs and midfielders. These results could be explained by time-motion characteristics of playing in different positions. Forwards and central defenders perform more match activities that depend on power performance - jumping, heading, kicking (Di Salvo et al., 2007). Young elite soccer players in late puberty are highly specialized both according to playing level and positional role on the field (Stroyer et al., 2004). Nevertheless, we cannot exclude the influence of a selection based explanation, more related to genetic factors. Differences in maximum force can derive not only from specific positional requirements but also from genetic factors (Bangsbo, 1994).

We did not find significant differences between positional roles in the agility test. In a study concerning talent detection and selection in soccer, agility proved to be the most powerful factor in discriminating between youth soccer players of elite and sub-elite levels (Reilly *et al.*, 2000). Thus, the performance level of the players (elite) can be a variable to consider for explaining these conflicting results. At the higher performance level, differences in physical performance between field positions seem to be of minor expression. However, others have reported that agility is one of the most important characteristics of the attackers during a soccer match (Gil *et al.*, 2007).

We did not find any significant differences in the results of the sprint tests, according to positional roles. However, in match-play at adult level the full-backs spend the highest percentage of time and cover the greatest distance in high-speed running and sprinting (Di Salvo *et al.*, 2007).

The intermittent endurance test is an indirect way of assessing the aerobic power of soccer players (Bangsbo, 1994). The differences between positions in results of Yo-Yo IE2 were not significant. In a soccer match, midfielders usually link the back and central parts of the field with the front line and, therefore, the midfield players have to run the longest distances in a match and need the highest endurance capacity (Di Salvo *et al.*, 2007; Gil *et al.*, 2007). The results contradict other studies indicating the midfield players as most demanding a good endurance capacity (Tumilty, 1993; Wisloff *et al.*, 1998). The lack of significant differences between groups could be related to the large standard deviations. An alternative explanation could be that intermittent endurance performance does not necessarily translate into a difference in performance between the various positions on the field in soccer (Bangsbo, 1994). Furthermore, trends observed in mature adult competitors may not apply to underage soccer players.

These findings reflect the general consensus within this state-based junior competition promoting overall player development, selecting those who are generally able to fulfil a range of positions rather than selecting position-specific players (Pyne *et al.*, 2006).

5. CONCLUSIONS

The purpose of this study was to describe and compare the performance in physical tests of youth soccer players according to positional roles. Although, each positional role has a different exercise intensity in a soccer match, no statistically significant difference was found between the different playing positions on any of the performance tests used in this study apart from the Squat Jump. A similar homogeneity is not anticipated in match-play characteristics.

Therefore, training prescriptions in soccer should be based on the specific requirements of each playing position to ensure that youth players are more able to fulfil their tactical responsibilities during football matches.

References

Bangsbo, J., 1994, The physiology of soccer - With special reference to intense intermittent exercise. *Acta Physiologica Scandinavica, Supplement,* **151**, pp. 1-155.

Bangsbo, J., Norregaard, L. and Thorso, F., 1991, Activity profile of competition soccer. *Canadian Journal of Sport Sciences,* **16**, pp. 110-116.

Bloomfield, J., Polman, R., Butterly, R. and O'Donoghue, P., 2005, Analysis of age, stature, body mass, BMI and quality of elite soccer players from 4 European Leagues. *Journal of Sports Medicine and Physical Fitness,* **45**, pp. 58-67.

Di Salvo, V., and Pigozzi, F., 1998, Physical training of football players based on their positional rules in the team. Effects on performance-related factors. *Journal of Sports Medicine and Physical Fitness,* **38**, pp. 294-297.

Di Salvo, V., Baron, R., Tschan, H., Calderon Montero, F. J., Bachl, N. and Pigozzi, F., 2007, Performance characteristics according to playing position in elite soccer. *International Journal of Sports Medicine,* **28**, pp. 222-227.

Gil, S. M., Gil, J., Ruiz, F., Irazusta, A. and Irazusta, J., 2007, Physiological and anthropometric characteristics of young soccer players according to their playing position: Relevance for the selection process. *Journal of Strength and Conditioning Research,* **21**, pp. 438-445.

Macarthur, D. G. and North, K. N., 2005, Genes and human elite athletic performance. *Human Genetics,* **116**, pp. 331-339.

Mohr, M., Krustrup, P. and Bangsbo, J., 2005, Fatigue in soccer: A brief review. *Journal of Sports Sciences,* **23**, pp. 593-599.

Pyne, D. B., Gardner, A. S., Sheehan, K. and Hopkins, W. G., 2006, Positional differences in fitness and anthropometric characteristics in Australian football. *Journal of Science and Medicine in Sport,* **9**, pp. 143-150.

Reilly, T., Bangsbo, J., and Franks, A., 2000, Anthropometric and physiological predispositions for elite soccer. *Journal of Sports Sciences,* **18**, pp. 669-683.

Reilly, T. and Thomas, V., 1976, A motion analysis of work rate in different positional roles in professional football match play. *Journal of Human Movement Studies,* **2**, pp. 87-97.

Rienzi, E., Drust, B., Reilly, T., Carter, J. E. and Martin, A., 2000, Investigation of anthropometric and work-rate profiles of elite South American international soccer players. *Journal of Sports Medicine and Physical Fitness,* **40**, pp. 162-169.

Stroyer, J., Hansen, L. and Klausen, K., 2004, Physiological profile and activity pattern of young soccer players during match play. *Medicine and Science in Sports and Exercise,* **36**, pp. 168-174.

Tumilty, D., 1993, Physiological characteristics of elite soccer players. *Sports Medicine,* **16**, pp. 80-96.

Vanderford, M. L., Meyers, M. C., Skelly, W. A., Stewart, C. C. and Hamilton, K. L., 2004, Physiological and sport-specific skill response of Olympic youth soccer athletes. *Journal of Strength and Conditioning Research,* **18**, pp. 334-342.

Wisloff, U., Helgerud, J. and Hoff, J., 1998, Strength and endurance of elite soccer players. *Medicine and Science in Sports and Exercise,* **30**, pp. 462-467.

CHAPTER THREE

Differences in technical skill performance of Portuguese junior soccer players according to competitive level and playing position

P. Vale, Á. Ramos, B. Salgado, P. Correia, P. Martins, J. Brito, E. Oliveira, A. Seabra and A. Rebelo

Faculty of Sports, University of Porto, Porto, Portugal

1. INTRODUCTION

Like their sporting heroes, many young athletes aspire to greatness, but only a smaller minority achieves greatness (Ward *et al.*, 2004; Vaeyens *et al.*, 2008). All over the world, a significant number of young boys dream to be a professional soccer player. Motivated by this dream, a large number of young soccer candidates look for the opportunity to be selected for the youth teams of soccer clubs.

Many organizations and top level teams, national federations and club teams, invest considerable resources in an effort to identify exceptionally gifted athletes. Abbott and Collins (2002) cited evidence that innate or pre-adolescent characteristics do not automatically translate into exceptional performance in adulthood. A number of factors such as maturation and training effect impact upon this development process.

Potentially talented athletes must be considered in a dynamic and multidimensional view (Abbott and Collins, 2002; Vaeyens *et al.*, 2008). In addition, in order to avoid exclusion of many 'promising' young players, the potential to develop must also be considered. Reilly *et al.* (2000) highlighted multifactorial characteristics of elite players and the fact that the requirements for soccer play can only be investigated using multivariate analysis, through the complex use of tests in an integrated form.

A well structured process of talent identification ('TID', i.e. the process of recognizing current participants with the potential to excel in a particular sport) and talent development ('TDE', i.e. providing the most appropriate learning environment to realize this potential), has a crucial role to play in the pursuit of excellence (Reilly *et al.*, 2000; Vaeyens *et al.*, 2008).

According to Malina *et al.* (2005), one of the main components involved in young athlete 'TID' and 'TDE' is sport-specific technical skill. Williams and Franks (1998), Williams and Reilly (2000) and Vaeyens *et al.* (2008) also referred to this component, emphasizing that in a majority of sports, soccer included, expertise or excellence may be achieved through different combinations of skills, sub-skills, attributes and capacities. In the particular case of soccer, technical skills evaluation acquires particular relevance, considering that soccer is a sport that requires refined skills for control, contact and passing or kicking the ball (Vaeyens *et al.*, 2006).

In view of the relevance of technical skills for match performance, more and specialized attention towards sport-specific skill assessment is necessary in young soccer players (Vanderford *et al.*, 2004; Malina *et al.*, 2005; Vaeyens *et al.*, 2006). These authors also underlined that a majority of investigations conducted with young soccer players concerned physical and anthropometric evaluations, like the studies of Garganta *et al.* (1993) and Hansen *et al.* (1999). In recent years, some studies focused attention on soccer technical skills and a variety of technical tests to evaluate different skills have been used - juggling, dribbling, passing and kicking are the more extensively used (Kirkendall *et al.* 1987; Van Rossum and Wijbenga, 1993; Rosch *et al.*, 2000; Seabra *et al.*, 2001; Malina *et al.*, 2005, 2007; Vaeyens *et al.*, 2006). Based on these studies, some results for different skill and age levels have been described. On the other hand, Ward *et al.* (2004) emphasized that according to deliberate practice theory (Helsen *et al.*, 1998), expertise results from the development of domain-specific knowledge structures and skills acquired through the process of adaptation to practice.

Unfortunately, information concerning young soccer players in the final stage of their formative process is rare. Talent development literature tends to situate this phase between the ages of 17 and 21 years (Ward *et al.*, 2004; Vaeyens *et al.*, 2008). In Portugal, '*juniores*' (17-19 years) is the ultimate competitive platform before young soccer players are confronted with the highest competitive level demands ('*senior*'). According to Ward and colleagues (2004), this is the stage for specialization.

Psychological and technical predictors are frequently ignored in 'TID' programmes (Abbott and Collins, 2004), despite the need to include these skills in any model of 'TID' (Vaeyens *et al.*, 2008). Specific positional requirements of players are a fundamental aspect to be taken into consideration. In some sports, component skills are not equally distributed across all playing positions but at the highest level, athletes must have a minimal competence level for each component.

Another important focus in 'TID' and 'TDE' themes is the idea that being at the highest level at the moment is as important as the ability to differentiate the athlete's potential for progression. It is important to find out which characteristics indicate that an individual has the potential to develop in sport and become a successful senior athlete. This question demands a very well defined selection - i.e., 'TIPS model' in AFC Ajax (Cohen, 1998) - or a longitudinal approach, such as the example presented by Vaeyens *et al.* (2008). They described a 'talent confirmation process' in UK Sport, a 3-6 month programme for individuals identified as talented, whereby athletes are confronted with the training requirements of elite sports competition. This exposure to systematic training is designed to support and validate the initial talent selection process. Vaeyens *et al.* (2008) suggested that researchers develop performance measures that better simulate the demands of actual competition, and more realistic test protocols in order to improve the predictive utility of the measures employed.

Talent identification processes require from coaches sufficient knowledge and preparation to define more relevant talent indicators (Vaeyens *et al.*, 2008). Given the time-consuming and financial needs associated with the long process of soccer player formation, the design of performance predictions for soccer players could represent a very valuable tool to economize on human and financial resources.

The aim of this study was to establish whether differences exist in technical skills performance between competitive level and playing positions of 'junior' (17-19 years) soccer players. In addition, we aimed to identify the technical tests that better discriminate the elite at national level from the non-elite at regional level.

2. METHODS

2.1 Participants

Eighty-seven junior sub-elite and 83 junior elite Portuguese soccer players participated in this study (Table 1). All sub-elite players were competing in the second division of a Regional Soccer League competition while the elite players were competing in the first division of a National Soccer League. The players were also grouped according to their playing position as follows: 26 full-backs (14 sub-elite and 12 elite), 27 central defenders (14 from the sub-elite group and 13 from the elite group), 73 midfielders (35 from the sub-elite and 38 elite) and 44 forwards (24 from the sub-elite group and 20 from the elite group).

Table 1. Mean values and standard deviations (SD) of age, height and body mass of the sample of Portuguese soccer players, according to competitive level and field position.

Participants				
Competitive level	N	Age (years)	Height (m)	Body mass (kg)
Sub-elite	87	17.91 ± 0.51	1.74 ± 0.06	67.90 ± 7.18
Elite	83	18.25 ± 0.52	1.76 ± 0.07	71.20 ± 10.05
Field positions				
Full-back	26	17.99 ± 0.60	1.74 ± 0.06	69.76 ± 6.12
Central Defender	27	18.04 ± 0.57	1.81 ± 0.05	74.83 ± 7.38
Midfielder	73	18.13 ± 0.49	1.74 ± 0.06	68.78 ± 7.40
Forward	44	18.07 ± 0.57	1.73 ± 0.06	67.30 ± 11.67

Defenders were separated into two groups (full-back and central defender), Midfielders (without wingers) were kept together and Forwards also included wingers. The criterion for this decision was based on the concept of Williams *et al.* (2008) of specific positional requirements and tasks. All players completed three different skill tests (juggling, dribbling and passing).

2.2 Procedures

Three tests of technical skills in soccer were administered to all players involved in the study. Two tests were adapted from the Ghent Youth Soccer Project (Vaeyens *et al.*, 2006) - *juggling and dribbling tests*. The other test - *passing test* - was adapted from the soccer skills proposed by the Portuguese Soccer Federation (F.P.F., 1986). Tests were administered outdoors on a playing field of artificial grass and the players performed a warm-up of twelve minutes duration composed of jogging and stretching exercises. The order of tests was the following: 1. Juggling; 2. Dribbling; and 3. Passing. The players wore soccer clothing and shoes. The official ball from the Portuguese Championships was used (Adidas Europass; ball size: 5). The pressure of the ball was maintained at 0.8 bar.

2.2.1 Juggling

The number of times the players touched the ball before it bounced on the ground was recorded. The juggling test (two trials) had a maximum score of 200 points (100 per attempt). If the ball hit the ground in the first two touches, the player re-started the test.

2.2.2 Dribbling

The players had to control the ball in a slalom dribble around nine cones (2 m apart) from the start to end lines and return. The objective was to complete the drill in the fastest time possible by controlling the ball only with the feet without knocking down the cones. If a cone was knocked over, the participant had to place it upright and continue the test. The average of the two values was used in the analysis.

2.2.3 Passing

The test of passing accuracy required the player to kick the ball into a goal divided in different marked sections. A standard official goal was used. The target was divided by ropes into six sections. One rope was placed horizontally between the posts at a height of 1.5 m. Two ropes were dropped from the crossbar, 0.5 m from each post. Five points were allocated for the upper right and left sections, and two points for the upper middle section. Three points were allocated for the lower right and left sections, and one point for the lower middle section. While standing outside the penalty area on the 18 yard line opposite the goal, the player had three attempts at kicking the ball into the goal. The maximum score was 15 points.

2.3 Statistical analysis

Initially, the Shapiro-Wilk test was used to test the normality of distribution of the variables studied (P>0.05). Descriptive statistics were calculated for all playing positions and competitive levels. Statistical analysis was performed using the independent student t-test and analysis of variance models. Before using these tests Levene's test was used to test homogeneity of variance (P>0.05). Scheffé test for multiple comparisons was used to check for specific differences due to playing positions. Statistical significance criterion was 0.05. SPSS 15.0 was used in all analyses.

3. RESULTS

Mean values and standard deviations (SD) of the results in skill tests and P values for independent measures t-tests are summarized in Table 2.

Table 2. Mean values, standard deviations (SD) and *P* values for independent measures, t-test for competition level.

	Sub-elite	Elite	
Soccer skills tests	Mean ± SD	Mean ± SD	p
Juggling (touches)	114.09 ± 61.52	144.30 ± 54.14*	0.04
Dribbling (s)	15.78 ± 1.42	15.42 ± 1.51	0.188
Passing (points)	6.39 ± 3.36	5.28 ± 2.60	0.49

* Significantly different from Sub-elite teams.

Elite youth soccer players performed a significantly higher number of touches of the ball in the juggling test than the sub-elite players (differences ± 30 touches). No significant differences were found between competitive level in passing and dribbling tests.

Table 3 shows the results of the analysis of variance model used to compare the mean values of the various skill tests according to field positions.

Focusing the analysis on field positions, midfield players showed significantly higher mean scores in the juggling test. In the dribbling test, again midfield players accomplished the test in a faster time than full-back players. No significant differences were found between playing positions in the passing test.

Table 3. Mean values, standard deviations (SD), F-test, and *P* values for analysis of variance models of different soccer skills test according to field position.

Tests	Full-back Mean ± SD	Central Defender Mean ± SD	Midfielder Mean ± SD	Forward Mean ± SD	F	P
Juggling	101.00 ± 58.51	127.65 ± 55.96	169.02 ± 38.36*	126.88 ± 58.56	10.51	<0.001
Dribbling	16.17 ± 1.13	15.24 ± 1.04	15.07 ± 1.36*	15.38 ± 1.57	3.24	0.024
Passing	6.44 ± 3.77	6.18 ± 2.89	5.69 ± 3.15	5.23 ± 2.59	0.66	0.57

* significantly different from the other field positions

4. DISCUSSION

Youth soccer players classified as elite and non-elite, or as being high and low in soccer ability, differ in soccer-specific skills (Hansen *et al.*, 1999; Rosch *et al.*, 2000). This observation was replicated in the present study, but only in part.

Analysing the performance of the junior soccer players in technical skills in relation to competitive level, significant differences were evident only in the juggling test: the elite soccer players performed a higher number of touches of the ball in juggling than the soccer players at sub-elite level. No significant differences were found between competitive level in passing and dribbling tests. According to Reilly *et al.* (2000) it was expected that elite players perform better at dribbling the ball than non-elite players. In view of the results of the Ghent Youth Soccer Project (Vaeyens *et al.*, 2006), in the dribbling test the existence of significant differences between the performance of soccer players at the highest competitive level (National) and at the lowest level (Regional) was anticipated.

Descriptive statistics from previous investigations by Malina *et al.* (2005) and Seabra *et al.* (2001) for Portuguese elite youth soccer players, of age range 13-15 years, showed a considerable variation in the scores, but none of the differences in the players grouped by position was significant. These results are similar to other samples reported in the literature (Vaeyens *et al.*, 2006; Malina *et al.*, 2007).

In the present study, we did not detect significant differences in passing skill between playing positions but the midfield players showed higher scores in juggling than the players in the other positions. In the dribbling test, the midfield players accomplished the test in a shorter time than the full-backs.

Williams *et al.* (2008) emphasized that component skills are not equally distributed across all playing positions. In fact, midfield players have to realize continuously a great number of technical actions - pass, control the ball, dribble - in small spaces and with great congestion of players. According to Vaeyens *et al.* (2006), they need to possess a good technical background in order to control the ball and become free themselves of adversaries, giving continuity to ball possession. These factors may also explain why the best results were obtained by midfielders in the juggling and dribbling tests.

At the final stage of the formation process, at least a minimal competence level for each component is required (Vaeyens *et al.*, 2006, 2008; Williams *et al.*,

2008). This fact may explain the lack of differences on the other tests, once technical differences may be less visible.

Results confirm that talent in soccer is more complex than a matter of skills tests administration. They corroborate the need to adopt a dynamic and multidimensional view, referred to by Reilly *et al.* (2000), Abbott and Collins (2002) and Vaeyens *et al.* (2008). In future investigations, it is desirable to work on the design of more sensitive skill tests leading to the technical discrimination of different competitive levels and more likely discrimination between elite and non-elite soccer players.

5. CONCLUSIONS

Given the importance of soccer and the amount of time and financial resources applied to soccer training programmes, talent identification and the ability to determine performance predictions for soccer players represent a very valuable tool to economize on human and financial resources. Furthermore, coaches must have sufficient knowledge to define more relevant talent indicators.

Results confirm that talent identification in soccer is more complex than simply test administration, and requires a dynamic and multidimensional approach. Differences in soccer-specific skills between youth soccer players classified as elite and non-elite, were also confirmed.

Results also provide evidence that juggling skill has a good potential to differentiate junior soccer players according to competitive level and that midfield players seem to have more refined skills than players of other positions, reflected in particular in the juggling and in the dribbling tests.

References

Abbott, A. and Collins, D., 2002, A theoretical and empirical analysis of a 'state of the art' talent identification model. *High Ability Studies*, **13**, pp. 157-78.

Abbott, A. and Collins, D., 2004, Eliminating the dichotomy between theory and practice in talent identification and development: considering the role of psychology. *Journal of Sports Sciences*, **22**, pp. 395-408.

Cohen, R., 1998, Futbol: deteccion y desarrollo del talento deportivo. *Educacion Fisica y Deportes*, **3**, pp. 10.

F.P.F., 1986, *Habilidades e Destrezas do Futebol: Os Skills do Futebol.* (Lisboa: Editora Federação Portuguesa de Futebol).

Garganta, J., Maia, J., Silva, R. and Natal, A., 1993, A comparative study of explosive leg strength in elite and non-elite young soccer players. In *Science and Football II*, edited by Reilly, T., Clarys, J. and Stibbe, A. (London: Routledge), pp. 304-306.

Hansen, L., Bangsbo, J., Twisk, J. and Klausen, K., 1999, Development of muscle strength in relation to training and level of testosterone in young male soccer players. *Journal of Applied Physiology*, **87**, pp. 1141-1147.

Helsen, W.F., Hodges, N.J. and Starkes, J.L., 1998, Team sports and the theory of deliberate practice. *Journal of Sport and Exercise Psychology*, **20**, pp. 12-34.

Kirkendall, D., Gruber, J. and Johnson, R., 1987, *Measurement and Evaluation for Physical Educators*, 2nd ed. (Illinois: Human Kinetics), pp. 123-125; 243.

Malina, R.M., Cumming, S.P., Kontos, A.P., Eisenmann, J., Ribeiro, B. and Aroso, J., 2005, Maturity-associated variation in sport-specific skills of youth soccer players aged 13-15 years. *Journal of Sports Sciences*, **23**, pp. 515-22.

Malina, R., Ribeiro, B., Aroso, J. and Cumming, S., 2007, Characteristics of youth soccer players aged 13-15 years classified by skill level. *British Journal of Sports Medicine*, **41**, pp. 290-295.

Reilly, T., 2006, Assessment of young soccer players: a holistic approach. *Perceptual and Motor Skills*, **103**, pp. 229-230.

Reilly, T., Williams, A., Nevill, A. and Franks, A., 2000, A multidisciplinary approach to talent identification in soccer. *Journal of Sports Sciences*, **18**, pp. 695-702.

Rosch, D., Hodgson, R., Peterson, L., Baumann, T.G., Junge, A., Chomiak, J. and Dvorak, J., 2000, Assessment and evaluation of football performance. *American Journal of Sports Medicine*, **28**, S29-S39.

Seabra, A., Maia, J.A. and Garganta, R., 2001, Crescimento, maturação, aptidão física, força explosiva e habilidades motoras específicas. Estudo em jovens futebolistas e não futebolistas do sexo masculino dos 12 aos 16 anos de idade. *Revista Portuguesa de Ciências do Desporto*, **1**, pp. 22-35.

Vaeyens, R., Lenoir, M., Williams, A.M. and Philippaerts, R., 2008, Talent identification and development programmes in sport: current models and future directions. *Sports Medicine*, **38**, pp. 703-714.

Vaeyens, R., Malina, R., Janssens, M., Renterghem, B., Bourgois, J., Vrijens, J. and Philippaerts, R., 2006, A multidisciplinary selection model for youth soccer: the Ghent Youth Soccer Project. *British Journal of Sports Medicine*, **40**, pp. 928-934.

Vanderford, M.L., Meyers, M.C., Skelly, W.A., Steward, C.C. and Hamilton, K.L., 2004, Physiological and sport-specific skill response of Olympic youth soccer athletes. *Journal of Strength and Conditioning Research*, **18**, pp. 334-342.

Van Rossum, J.H.A. and Wijbenga, D., 1993, Soccer skills technique tests for youth players: construction and implications. In *Science and Football II*, edited by Reilly, T., Clarys, J. and Stibbe, A. (London: Routledge), pp. 313-318.

Ward, P., Hodges, N.J., Williams, A.M. and Starkes, J., 2004, Deliberate practice and expert performance defining the path to excellence. In *Skill Acquisition in Sport Research, Theory and Practice*, edited by Williams, A.M. and Hodges, N.J. (London: Routledge), pp. 231-258.

Williams, A.M. and Franks A., 1998, Talent identification in soccer. *Sports Exercise and Injury*, **4**, pp. 159-65.

Williams, A.M. and Reilly T., 2000, Talent identification in soccer. *Journal of Sports Sciences*, **18**, pp. 657–67.

Williams, A.M., Ward, L., Julian, D. and Smeeton, N.J., 2008, Domain specificity, task specificity and expert performance. *Research Quarterly for Exercise and Sport*, **79**, pp. 428-433.

CHAPTER FOUR

The mechanisms underpinning decision-making in youth soccer players: an analysis of verbal reports

R. Vaeyens[1], M. Lenoir[1], A.M. Williams[2], S. Matthys[1] and R.M. Philippaerts[1]

[1]Department of Movement and Sports Sciences, Ghent University, Ghent, Belgium
[2]Research Institute for Sport and Exercise Sciences, Liverpool John Moores University, Liverpool, UK

1. INTRODUCTION

An awareness that perceptual-cognitive skill is essential for proficient behaviour in sport has initiated scientific interest in examining the mechanisms underpinning successful performance (Savelsbergh *et al.*, 2005). There is a consensus that elite athletes demonstrate superior perceptual-cognitive skill when compared with less elite counterparts (Williams and Ward, 2007). Although the skilled performer's superior performance is underpinned by a number of perceptual-cognitive skills that are seamlessly integrated during task performance, researchers have merely focused on identifying differences in performance between skilled and less skilled performers rather than revealing the mechanisms underlying expert performance (Williams and Ericsson, 2005). However, an increasing number of scholars have recently shown interest in explaining expert performance and describing the mediating perceptual-cognitive processes that contribute to expertise. Several types of process-tracing measures and task manipulations have been suggested in applied domains to examine the mechanisms underlying perceptual-cognitive expertise (Williams and Ericsson, 2005; Hodges *et al.*, 2007). In sport, researchers have primarily employed eye-movement registration techniques during task performance (Williams *et al.*, 1999; Starkes *et al.* 2001). However, the use of eye movement registration has specific limitations (Wright and Ward, 1994; Zelinsky *et al.*, 1997) and several authors have argued that the manner in which the information is processed and evaluated is at least as important as the way in which it is extracted from the display (Abernethy and Russell, 1987).

Whilst most researchers have focused on identifying the visual search behaviours underlying successful performance, few have systematically examined how this information is processed to make a decision (for exceptions see McPherson, 1999a; 1999b). This issue may be addressed by employing verbal report techniques to elicit the nature of cognitions used during representative task performance. This work originates from the information-processing framework (see Newell and Simon, 1972; Ericsson and Simon, 1993). The rationale of collecting verbal reports is to extend knowledge on the memory structure of skilled performers and how information is stored, retrieved and processed. Due to the severe time constraints apparent in many sport settings, performers need to possess skills that bypass the customary limitations imposed on short-term working memory. Simon and Chase (1973) initially put forward the 'chunking' theory to account for these effects, whereas, more recently, Ericsson and colleagues (1993) advocated that performers develop memory skills that promote both rapid encoding of information in long-term memory and allow selective access to this information when required. The long-term working memory theory proposed by Ericsson and Kintsch (1995) suggests that complex retrieval structures in the long-term memory are accessible by means of cues in short-term memory.

In the present study we intended to gain more insight into how information is being processed prior to successful decision-making skill in youth soccer. We employed a retrospective verbal report procedure during a perceptual-cognitive 'game-reading' task to examine whether there are differences in thought processes in a group of youth soccer players of comparable experience level. In line with previous findings, it was hypothesised that elite players demonstrate superior decision-making skill when compared to non-elite players. As previous verbal protocol studies have revealed that experts have greater declarative, procedural and strategic sport-specific knowledge than novices (Starkes *et al.* 2001), we expected to observe differences between elite and non-elite youth soccer players.

2. METHODS

2.1 Participants

A total of 17 skilled and 14 less skilled male youth soccer players provided informed parental consent prior to the start of the study. Skilled players (*M* age = 15.3 ± 0.5 years) currently attended one of the five elite soccer academies in Flanders and played at the highest youth team level. The majority of elite participants had represented the national youth team in Belgium. Non-elite players (*M* age = 15.3 ± 0.5 years) played at recreational level in regional teams. Both groups started playing soccer at a similar age. The study was approved by the Ethics Committee of the Ghent University Hospital.

2.2 Test film

Decision-making skill was examined using film-based simulations of offensive patterns of play in soccer (see Vaeyens *et al.*, 2007a, b). Participants were required to imagine themselves as an offensive midfielder playing in a central position just in front of the camera. The simulations varied in the number of players presented on film: 2 *vs.* 2, 3 *vs.* 1, 3 *vs.* 2, 4 *vs.* 3, and 5 *vs.* 3. Each condition also included a goalkeeper. A sequence of play lasted around six seconds (range 3.0-9.7 s) and included an offensive pattern of play with variations in the positions of defenders and attackers. Each sequence ended with a pass towards the central offensive midfield player. The film sequences were occluded shortly after the ball was played towards this player to avoid participants from receiving any further information.

2.3 Apparatus

Soccer-specific film simulations, movement-based response measures and verbal report techniques were used to assess decision-making processes and skill. To increase ecological validity, film clips were projected using an LCD projector (NEC VT 670-XGA, Hong Kong) on a 4.3 m x 2.5 m test screen, positioned 4.4 m from the participant. Pressure sensitive mats located underneath each participant's feet and the ball, positioned a distance of 1 m in front of the participant, recorded the decision times. During the task response accuracy and retrospective verbal reports were collected via a digital voice recorder (Olympus DS-2200).

2.4 Procedure

Prior to the experimental task, participants received training on how to think aloud and provide verbal reports. The instructions employed were based on the protocol of Ericsson and Simon (1993). This training session comprised both instruction and practise on thinking aloud and giving retrospective verbal reports in solving general and soccer-specific questions. On average, the verbal report training protocol lasted approximately 40 minutes. On completion of the instruction phase, a standard explanation of the test procedure was provided in which participants were told that they would view a number of offensive soccer simulations. The task was to make the correct decision quickly and accurately once the ball was played in the direction of the central offensive midfielder (i.e., the player representing themselves on screen). To increase the realism of the experimental setting, participants responded either by (1) passing the ball to a player on screen, (2) shooting the ball towards goal, or (3) moving as if to dribble the ball around a defender. Additionally, to ensure that the intended technical skill was carried out, participants were required to verbalise their intended response immediately after each trial and to provide retrospective verbal reports after every second trial.

At the end of the instruction phase participants were presented with four practise trials to ensure familiarisation with the experimental setting and the verbal protocol procedure. A total of 18 offensive patterns were then presented: two 2 *vs.* 2, four 3 *vs.* 1, four 3 *vs.* 2, four 4 *vs.* 3, and four 5 *vs.* 3 situations. The order of presentation of film clips was randomised for all participants. An inter-trial interval of approximately 30 s was employed and the entire test session was completed in around 25 min.

2.5 Dependent Variables and Data Analysis

Two performance measures were recorded on the decision-making skill test (Vaeyens *et al.*, 2007a, b). The *decision time (DT)* was the time from the start of the final pass towards the central offensive midfield player to the initiation of the participant's movement response (in ms). The *response accuracy (RA)* was calculated as the mean correctness of the participant's response relative to the pre-labelled codes across all simulations (in %).

In correspondence to the methodology applied by North (2007), participants' retrospective verbal reports were transcribed verbatim. They were subsequently coded on two separate classification schemes, namely types of action and types of stimuli.

In general, *actions* were verbs that express behaviours or specified types of play (e.g. cross, dribble). Participants verbalised a total of 121 different action statements. Next, similar action statements were grouped in 15 distinct types of action categories: shots at goal, crosses, passing across the defence, short passes, dribbles, controlling the ball, unmarked positioning, movement into central space, movement into wide space, attacking runs, supporting player with ball, creating space, defensive marking, pressuring and defending, and visual actions.

Stimuli were features within the display to which the participants referred. The 42 types of stimuli participants mentioned were grouped into 8 similar references: defensive team, ball and player with ball, central offensive midfielder, attacking team, goal, free space, space and situation.

The performance measurements and the total number of actions or stimuli verbalised were analysed using independent-samples t tests. In order to determine whether the groups reported different types of actions or stimuli, separate MANOVAs were employed with group as the fixed factor and the 15 action and 8 stimuli categories respectively as the dependent variables. Post-hoc Bonferroni corrected comparisons were used to identify potential group differences.

3. RESULTS

3.1 Performance data

The non-elite players (1017.79 ± 315.65 ms) responded significantly ($t_{19.4} = 4.4$, P < 0.001) slower than their elite counterparts (599.95 ± 174.24 ms). The two groups did not differ in response accuracy ($t_{29} = -0.1$, P $= 0.93$; elite, 82.2 ± 5.4 %; non-elite, 82.0 ± 8.8%).

3.2 Verbal reports

Actions. The elite (40.24 ± 9.50) and non-elite (37.86 ± 12.78) groups did not differ in the number of verbalised action statements ($t_{29} = 0.4$, P $= 0.56$). MANOVA revealed no significant group effect, $F_{15,15} = 0.79$, P $= 0.67$. Due to the specific task (choosing one of three response alternatives), most verbalisations were made to short passes, shots at goal and dribbles (Figure 1).

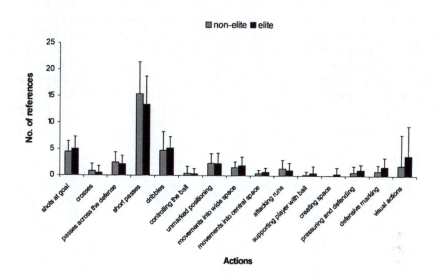

Figure 1. Mean (+ s) number of actions verbalised by elite and non-elite participants.

Stimuli. The elite (47.29 ± 16.84) and non-elite (42.93 ± 19.37) groups did not differ in the number of stimuli statements (t_{29} = -0.7, P = 0.51). MANOVA revealed no significant group effect, $F_{8,22}$ = 1.14, P = 0.38. However, Bonferroni corrected pairwise comparisons showed that the elite group made significantly more references to free space and analysis of the situation than the non-elite group. Figure 2 shows that most verbalisations were made to space, attacking team and player in possession of the ball.

4. DISCUSSION

We examined differences in decision-making skill and underlying thought processes between elite and non-elite youth soccer players using simulations of offensive play in soccer. The focus of this experiment was to elicit thought processes that reflect the nature of expertise during representative task performance via the use of retrospective verbal reports. In line with previous research, elite participants were expected to perform significantly better than sub-elite players due to superior encoding and memory skills.

The superior performance of elite players in this study is consistent with the findings from Vaeyens *et al.* (2007a), confirming the validity of the laboratory decision-making skill test. The verbal report data revealed fewer group differences than expected. In contrast to earlier findings (North, 2007), in the current study both groups did not differ in the number of stimuli and action statements. We could only observe significant group differences for the number of references to free space and the situation. The notion that elite players more frequently reported these stimuli may reflect their more analytical and goal-oriented thought processes. However, contrary to our expectations, the present data could provide no clear evidence that elite youth soccer players have developed more complex memory structures than non-elite counterparts. Although the verbal report technique has been suggested as the standard procedure to elicit the nature of cognitions during representative task performance (Williams and Ericsson, 2005), experiences from the current experiment lead us to raise the question if this method is appropriate for each type of experiment.

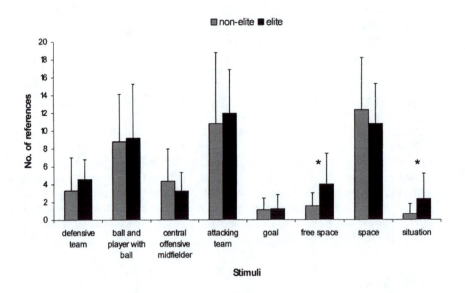

Figure 2. Mean (+ s) number of stimuli verbalised by elite and non-elite participants.

We consider several reasons that may account for the unsuccessful application of this procedure in the current experiment. First, the participants in this study were adolescents whereas previously researchers have mainly examined adult athletes (e.g., Williams and Davids, 1998; McPherson, 1999b, 2000; North, 2007). It may be hypothesised that youngsters lack the ability to verbalise their thoughts clearly. This notion was supported by the extreme low number of verbalisations that limited further analysis of the available data. When comparing verbal report data from adult and youth tennis players, McPherson (1999a) concluded that adults accessed more sophisticated problem representations than youth. One could argue that the verbal report protocol may not be equally suited for adolescents as it is assumed to be for adults. Moreover, soccer players are not used to expressing their thoughts during the game. In a personal post-experiment interview the participants confirmed the aforementioned notions and reported difficulties in expressing their thoughts during the test. A third argument could be related to the specific nature of the task. In the present experiment the participants were required to respond as quickly as possible, which may have negatively affected their ability to verbalise their thoughts. Other researchers have circumvented this problem by eliciting verbal reports during tasks without time restrictions either in real life situations (e.g., McPherson, 1999a, 1999b, 2000) or during simulated situations (e.g., Williams and Davids, 1998; North, 2007).

In conclusion, the current data could not offer concrete and indisputable evidence that experts have greater declarative, procedural and strategic sport-specific knowledge than less skilled players and highlight potential problems when employing the verbal report technique. We argue that the verbal protocol technique may not be universally appropriate. Alternative methodologies that can identify the mechanisms mediating superior tactical skill may therefore be helpful.

References

Abernethy, B. and Russel, D.G., 1987, The relationship between expertise and visual search strategy in a racquet sport. *Human Movement Science,* **6**, pp. 283- 319.

Ericsson, K.A. and Delaney, P.F., 1999, Long-term working memory as an alternative to capacity models of working memory in everyday skilled performance. In *Models of Working Memory: Mechanisms of Active Maintenance and Executive Control,* edited by Miyake, A. and Shah, P. (Cambridge, UK: Cambridge University Press), pp. 257-297.

Ericsson, K.A. and Kintsch, W., 1995, Long-term working memory. *Psychological Review,* **102**, pp. 211-245.

Ericsson, K.A., Krampe, R.T. and Tesch-Römer, C., 1993, The role of deliberate practice in the acquisition of expert performance. *Psychological Review,* **100**, pp. 363-406.

Ericsson, K.A. and Simon, H.A., 1993, *Protocol Analysis: Verbal Reports as Data* (Cambridge, MA: MIT Press).

Hodges, N.J., Huys, R. and Starkes, J.L., 2007, Methodological review and evaluation of research in expert performance in sport. In *Handbook of Sport Psychology,* 3rd ed., edited by Tenenbaum, G. and Eklund, R.C. (Hoboken, NJ: Wiley), pp. 161-183.

McPherson, S.L., 1999a, Expert-novice differences in performance skills and problem representations of youth and adults during tennis competition. *Research Quarterly for Exercise and Sport,* **70**, pp. 233-251.

McPherson, S.L., 1999b, Tactical differences in problem representations and solutions in collegiate varsity and beginner women tennis players. *Research Quarterly for Exercise and Sport,* **70**, pp. 369-384.

McPherson, S.L., 2000, Expert-novice differences in planning strategies during collegiate singles tennis competition. *Journal of Sport and Exercise Psychology,* **22**, pp. 39-62.

Newell, A. and Simon, H.A., 1972. *Human Problem Solving* (Englewood Cliffs, NJ: Prentice-Hall).

North, J., 2007, The mechanisms underlying skilled anticipation and recognition in a dynamic and temporally constrained domain. Liverpool John Moores University, Doctoral dissertation, pp. 72-111.

Savelsbergh, G.J.P., Williams, A.M., van der Kamp, J. and Ward, P., 2005, Anticipation and visual search behaviour in expert goalkeepers. *Ergonomics,* **48**, pp. 1686-1697.

Simon, H.A. and Chase, W.G., 1973, Skill in chess. *American Science,* **61**, pp. 394-403.

Starkes, J.L., Helsen, W.F. and Jack, R., 2001, Expert performance in sport and dance. In *Handbook of Research in Sport Psychology,* edited by Singer, R.N., Hausenblas, H.A. and Janelle, C.M. (New York: John Wiley), pp. 174-201.

Vaeyens, R., Lenoir, M., Williams, A.M., Mazyn, L. and Philippaerts, R.M., 2007a, The effects of task constraints on visual search behavior and decision-making skill in youth soccer players. *Journal of Sport and Exercise Psychology,* **29**, pp. 147-169.

Vaeyens, R., Lenoir, M., Williams, A.M. and Philippaerts, R.M., 2007b, The mechanisms underpinning successful decision-making in skilled youth soccer players: Analysis of visual search behaviors. *Journal of Motor Behavior,* **39**, pp. 395-408.

Ward, P. and Williams, A.M., 2003, Perceptual and cognitive skill development in soccer: The multidimensional nature of expert performance. *Journal of Sport and Exercise Psychology,* **25**, pp. 93-111.

Williams, A.M. and Davids, K., 1998, Visual search strategy, selective attention, and expertise in soccer. *Research Quarterly for Exercise and Sport,* **69**, pp. 111-128.

Williams, A.M., Davids, K. and Williams, J.G., 1999, *Visual Perception and Action in Sport* (London: E and FN Spon).

Williams, A.M. and Ericsson, K.A., 2005, Perceptual-cognitive expertise in sport: Some considerations when applying the expert performance approach. *Human Movement Science,* **24**, pp. 283-307.

Williams, A.M. and Ward, P., 2007, Anticipation and decision making: Exploring new horizons. In *Handbook of Sport Psychology,* 3rd ed., edited by Tenenbaum, G. and Eklund, R.C. (Hoboken, NJ: Wiley), pp. 203-223.

Wright, R.D. and Ward, L.M., 1994, Shifts of visual attention: An historical and methodological overview. *Canadian Journal of Experimental Psychology,* **48**, pp. 151-166.

Zelinsky, G.J., Rao, R.P.N., Hayhoe, M.M. and Ballard, D.H., 1997, Eye movements reveal the spatiotemporal dynamics of visual search. *Psychological Science,* **8**, pp. 448-453.

CHAPTER FIVE

Youth development in elite European football: structure, philosophy, and working practices

H. Relvas[1], D. Richardson[1], D. Gilbourne[2] and M. Littlewood[1]

[1] Research Institute for Sport and Exercise Sciences,
Liverpool John Moores University, Liverpool, UK
[2] University of Wales, Cardiff

1. INTRODUCTION

Football (soccer) captures the attention of millions of people around the world (Parker, 1995). The public interest together with the commercialism around the game has led professional clubs to operate as service enterprises engaged in the business of performance, entertainment, and with the aim of acquiring financial profit (Bourke, 2003; Vaeyens *et al.*, 2005).

Due to this business orientation, professional soccer clubs tend to invest in more high-profile players who are more likely to improve merchandise sales and match results. This investment in more 'finished' players, suggests a lack of readiness and/or even willingness to prepare young talented players for the first team environment. To change the perceived reluctance of professional soccer clubs to invest in youth development programmes (Richardson *et al.*, 2005), and due to the apparent lack of emerging young talent in the Union of European Football Associations (UEFA, 2005), measures to restrict player purchases and encourage youth development have been introduced by national and international football associations. These measures, and the increased values of young players, were designed to encourage professional soccer clubs to invest in youth academies, talent identification and development processes (Williams and Reilly, 2000; Reilly *et al.*, 2003; Vaeyens *et al.*, 2005).

With the increasing number of professional clubs operating as business enterprises, it would appear essential to reduce the risk of the investment (i.e., both financial and time-intensive investments) in youth training programmes (Gonçalves, 2003). This report extends the work developed by Relvas and colleagues (2009), that aimed to explore the different organisational structures, philosophies and working practices concerning youth development within elite European professional soccer clubs. The sentiments and context of Relvas and colleagues' (2009) findings are drawn on throughout this report and extended with the introduction of two more clubs from another country (i.e., France).

1.1 Complexity of youth talent development

According to Stratton and colleagues (2004, p.183), '...talent development is predominantly associated with the provision of a suitable environment from which potential talent can be realised.' Talent development incorporates social, intellectual, educational, welfare, physiological, physical, and psychological factors. Development itself depends on several elements – the efficiency of the sport organization; the sport's human resources; methods of coaching and training; and the application of sports medicine and sports sciences (Maguire and Pearton, 2000). When a player enters a systematic developmental process, the objective is to develop playing ability and nurture the individuals towards realising their potential (Reilly *et al.*, 2000).

The general mission of the football academies or youth departments is to develop elite players for the first team, not only to increase the quality of the squad but to also generate income through player sales (Richardson *et al.*, 2004; Laurin *et al.*, 2008). In this sense, a more structured approach to developing talent may reap both sporting and financial rewards (Stratton *et al.*, 2004). The organisational structure, associated working practices and culture of any given organisation will ultimately influence the performance and success of the human resource (e.g., players). Indeed, Durand-Bush and Salmela (2001, p.285) stated that '...we cannot change our genetic makeup, but we can change our environment to make it as conducive as possible to improving performance...'

The pursuit of financial profit has also extended to the team squads and players. The pressure and 'need to win' (at all costs) culture requires clubs to invest in more (experienced) high-profile players. Managers give priority to more experienced players that they believe could give them more chance of instant success (Maguire and Pearton, 2000). In this regard, players who are more ready and/or equipped to cope with the demands of first team soccer are favoured. Such a position would appear to contribute to a lack of emerging young talent in the UEFA Federations. Subsequently, measures to rectify this position have been explored by national and international confederations. For example, in England, the implementation of the 'Charter for Quality' by the English Football Association (Football Association Technical Department, 1997), focused a concern on the young player's development (Stratton *et al.*, 2004); in France, the 'Charte du football professionnel' (2007) is a document updated and released every season to regulate the Academies of French professional soccer clubs. Similarly, UEFA established that, by the 2008-09 season, each club should include in its squad four players from its own academy and four others from clubs of the same national association (UEFA, 2005). However, the EU legislation (i.e., freedom of movement) prevents UEFA from restricting player movement within the European Union. Consequently (and more worryingly) clubs have been undertaking a more global strategy for recruitment of young players.

1.2 Coherent environment for talent development

If the aim of youth academies is to develop and promote youth players to the first team and/or professional environment, it seems pertinent to explore the later years of development. This final period of development between 17 and 21 years of age has been considered as a critical period in the process of development for the player's future career (Richardson *et al.*, 2005; Vaeyens *et al.*, 2005). This critical period is associated with a high proportion of drop-outs, conflicts of adulthood, difficulties of peer group acceptance (i.e., young players start to interact with the players from the reserves and first team), and inadequate adaptability to the professional football environment (Parker, 2001). At a competitive level, the transition to the first team is associated with a higher level of performance expectation alongside a reduced tolerance for failure. Furthermore, players may experience, and adjust to, new and/or distinct organisational sub-cultures that exist within the professional environment.

A successful transition here may also depend on external factors such as, the opportunities to play, the absence of injuries, the nature of guidance and training, and personal, social, and cultural factors (Reilly *et al.*, 2000). It has been suggested that young players would benefit from a more structured, coherent and informed development approach in order to aid a player's, practitioner's and/or club's understanding of the complexity of such a critical transition (Richardson *et al.*, 2005).

It appears pertinent to explore the determinants of a suitable and coherent environment, which may facilitate the process of development and help players progress to a higher level (Relvas *et al.*, 2009).

2. METHODS

Twenty-six Head of Youth Development (HYD), Academy Managers (AM) or equivalent from top-level clubs across five European countries, namely England (n=6), France (n=2), Portugal (n=5), Spain (n=9), and Sweden (n=4) were interviewed. Twenty of the top-level clubs (i.e., currently playing in the top league of their respective country at time of the interview) were involved in European competitions. The interviews followed a semi-structured interview schedule (Biddle *et al.*, 2001) that was deductively developed through pre-determined conceptual themes, and with respect to informal contact with a selection of youth development staff within professional soccer.

Face-to-face interviews were transcribed verbatim and (predominantly) deductively analysed using content analysis procedures (Scanlan *et al.*, 1989; Côté *et al.*, 1993). Due to the vast amount of data collected, the data were grouped according to the themes defined in the interview schedule and other, more inductive, themes that emerged from the interviews. Emerging themes were subjected to within-method triangulation (McFee, 1992).

Strategic, operational policy and practice documentation for all clubs was collated from a range of secondary sources. The material was obtained directly from some staff personnel and/or through the club's internet site.

3. RESULTS AND DISCUSSION

Even given the introduction of two new clubs from a new country to the analysis (see Relvas and colleagues, 2009), it was evident that the clubs operated within one of the two distinct "general" organisational structures. Structure A (n=18) identified a 'sports director' who appeared to operate as a link between the Board and each of the distinct football departments (i.e., youth and professional). Structure B (n=8) also identified an Executive Board but it appeared to be (more) directly responsible for both the youth and professional environments. Neither of these distinct organisational structures could be associated with any one particular country or culture (see structures a. and b. respectively in Figure 1).

The first team manager was identified as the direct line manager to the youth department (i.e., HYD or AM) in only two clubs (one English and one French). In one of the clubs with structure B, it was possible to find business managers controlling the professional and youth departments. Each one of these business managers controls and rules these distinct departments as a small enterprise looking not only for athletic performance but also financial profit.

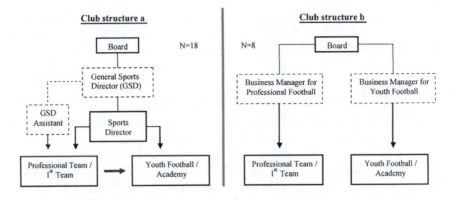

Figure 1. Representation of the two types of club structure evidenced within the 26 clubs across 5 European countries.

Two distinct youth development structures were also evidenced. The majority of the clubs appeared to favour the identification of different departments (e.g. technical, medical) similar to that outlined by Stratton *et al.* (2004). The Swedish clubs operated with age group personnel teams (i.e., each age group team operated with the same relevant personnel). Each club identified 'similar' youth staff positions; however, the interviews with the HYD suggested that their specific roles and responsibilities appeared to vary (slightly) within each club. The different roles and responsibilities of the youth staff members might be affected by the specificity of each club environment, although to keep the effectiveness of their work, it seems necessary to define clearly the organisation structure, responsibility, and authority (Rogers *et al.*, 1994). In the majority of the clubs there appeared to be no documentation available clarifying exactly what is expected from each one of the youth staff members. Such practices might lead to role ambiguity in their daily practice.

It was clear that the predominant aim of each youth development programme was to develop players for the first team. However, all clubs recognised other benefits of the development process, such as the player's personal development, and financial reward. Similarly, Stratton *et al.* (2004, p.201) stated that '… academies aspire to develop players for the first team or (at least) generate income through the sale of 'marketable assets'…', and also develop the 'whole' individual. Only the Swedish clubs offered a sense that their purpose was to also develop players for the Swedish National side.

The predominant aim was to develop players for the first team, so it seems pertinent to analyse how the clubs establish the communication between the youth and the first team. It was possible to identify three different practices: a) one where the older youth team players use the same training facilities as the first team (N=16), although this did not mean that they trained at the same time; b) where the contact between the youth and professional environment was co-ordinated through a sports director with no apparent 'direct' contact between the HYD and the first team manager (N=12); and in some clubs, it appeared that there was no regular contact between the first team and the youth environment (N=10). In some cases the first team and youth environments existed in different geographical locations, contributing to a formal 'distance' between the two departments. The perceived lack of communication and/or the apparent inconsistency of the contacts (i.e., between the youth and first team environments), particularly in the latter scenario, seemed to contribute to the appearance of different 'game' cultures (e.g., attacking versus defensive strategies, a passing style versus a long ball style) within the same club, and consequently dissatisfaction within the staff members.

4. CONCLUSION and FUTURE DIRECTIONS

The inclusion of another European country (i.e., through two elite French soccer clubs) further confirms the findings of Relvas and colleagues (2009). In particular, specific philosophies, structures and/or working practices towards youth development, were not peculiar to any particular country. Different organisational structures were evident alongside more subtle working practice distinctions between clubs. For example, distinct approaches to the identification of the specific role and responsibility of the practitioners, the presence, function and operationalisation of the reserves (or 'B' teams), the pragmatics of the transition from youth to the professional team, communication mechanisms (e.g., first team/ youth environment), and the dominant presence of a more club orientation towards the development of young players (i.e., only in Sweden did they offer a more national side orientation) were evidenced. Notably, the apparent non-involvement of the first team manager with the youth team and subsequent lack of formal proximity and communication between the youth and first teams, regardless of the club's structure, tended to hinder the coherent progression of young players into the first team environment.

Future work should seek to explore the working practices of the youth development practitioners themselves, not only to understand the operationalisation of the practitioner's working mechanisms, roles, and responsibilities, but also to clarify some concepts and development practices not completely explained or articulated by the HYDs. Further research should attend to the daily practices and experiences lived by young players (also see Helsen *et al.*, 2000; Volossovitch, 2003). This present report provides further understanding of the youth football development environment, its culture and its characteristics, which may provide youth practitioners with tools to improve the preparation of players for the critical transition from the youth to the professional football environment.

References

Biddle, S., Markland, D., Gilbourne, D., Chatzisarantis, N. and Sparkes, A., 2001, Research methods in sport and exercise psychology: quantitative and qualitative issues. *Journal of Sports Sciences*, **19**, pp. 777-809.

Bourke, A., 2003, The dream of being a professional soccer player: insight on career development options of young Irish players. *Journal of Sport and Social Issues*, November, **27**, pp. 399-419.

Côté, J., Salmela, J.H., Baria, A. and Russell, S., 1993, Organising and interpreting unstructured qualitative data. *The Sport Psychologist*, **7**, 127-137.

Durand-Bush, N. and Salmela, J., 2001, The development of talent in sport. In *Handbook of Sport Psychology*, 2nd ed., edited by Singer, R., Hausenblas H. and Janelle, C. (USA: John Wiley & Sons, Inc), pp. 269-289.

Football Association Technical Department, 1997, *Football Education for Young Players: "A Charter for Quality"* (London: The Football Association).

Gonçalves, C., 2003, Fórum – Selecção e detecção de talentos. *Treino desportivo*, Abril, Ano V, **21 – 3ª série**, pp. 32-33.

Helsen, W.F., Hodges, N.J., Van Winckel, J. and Starkes, J.L., 2000, The roles of talent, physical precocity and practice in the development of soccer expertise. *Journal of Sports Sciences*, **18**, pp. 727-736.

Laurin, R., Nicolas, M. and Lacassagne, M., 2008, Effects of a personal goal management program on school and football self-determination motivation and satisfaction of newcomers within a football training centre. *European Sport Management Quarterly*, **8**, pp. 83-99.

Ligue Professionnel Football, 2007, Charte du Football Professionnel – Saison 2007/2008.http://www.lfp.fr/reglements/charteFootballProfessionnel.asp (accessed 11/06/2008)

Maguire, J. and Pearton, R., 2000, The impact of elite labour migration on the identification, selection and development of European soccer players. *Journal of Sports Sciences*, **18**, pp. 759-769.

McFee, G., 1992, Triangulation in research: Two confusions. *Educational Research*, **34**, pp. 215-219.

Parker, A., 1995, Great expectations: grimness or glamour? The football apprentice in the 1990s. *The Sports Historian*, **15**, 107-126.

Parker, A., 2001, Soccer, servitude and sub-cultural identity: football traineeship and masculine construction. *Soccer and Society*, **2**, pp. 59-80.

Reilly, T., Williams, A.M., Nevill, A. and Franks, A., 2000, A multidisciplinary approach to talent identification in soccer. *Journal of Sports Sciences*, **18**, pp. 695-702.

Reilly, T., Williams, A.M. and Richardson, D., 2003, Identifying talented players. In *Science and Soccer*, edited by Reilly, T. and Williams, A.M. (Routledge: London), pp. 307-326.

Relvas, H., Richardson, D., Gilbourne, D. and Littlewood, M., 2009, Youth development structures, philosophy and working mechanisms of top-level football clubs: a Pan European perspective. In *Science and Football VI – The Proceedings of the Sixth World Congress on Science and Football*, edited by Reilly, T. and Korkusuz, F. (New York: Routledge), pp. 476-481.

Richardson, D, Littlewood, M. and Gilbourne, D., 2004, Developing support mechanisms for elite young players in a professional soccer Academy: creative reflections in action research. *European Sport Management Quarterly*, **4**, pp. 195-214

Richardson, D., Littlewood, M. and Gilbourne, D., 2005, Homegrown or home Nationals? Some considerations on the local training debate. *Insight Live* https://ice.thefa.com/ice/livelink.exe/fetch/2000/10647/466509/477135/4772 57/Homegrown_or_Home_Nationals._The_Case_for_the_Local_Training_ Debate.?nodeid=675785&vernum=0 (accessed 20/09/2005).

Rogers, R., Li, E. and Ellis, R., 1994, Perceptions of organizational stress among female executives in the U.S. government: An exploratory study. *Public Personnel Management,* **23**, pp. 593–609.

Scanlan, T.K., Ravizza, K. and Stein, G.L., 1989, An in-depth study of former elite figure skaters: I. Introduction to the project. *Journal of Sport and Exercise Psychology*, **11**, pp. 54-64.

Stratton, G., Reilly, T., Williams, A.M. and Richardson, D., 2004, *Youth Soccer – From Science to Performance* (London: Routledge).

UEFA, 2005, Formação recebe luz verde. http://pt.uefa.com/news/newsId=297234 (accessed 25/04/2005).

Vaeyens, R., Coutts, A. and Philippaerts, R., 2005, Evaluation of the "under-21 rule": Do young adult soccer players benefit? *Journal of Sports Sciences*, **23**, pp. 1003-1012.

Volossovitch, A., 2003, Fórum – Selecção e detecção de talentos. *Treino desportivo*. Abril, Ano V, **21 – 3ª série**, pp. 34-35.

Williams, A.M. and Reilly, T., 2000, Talent identification and development in soccer. *Journal of Sports Sciences*, **18**, pp. 657-667.

Part II

Biomechanics and

Research Methods

CHAPTER SIX

Knee flexion and ankle extension strategies at instep penalty kick

A. Goktepe[1], E. Ak[3], H. Karabork[2], S. Cicek[3] and F. Korkusuz[3]

[1] Technical Science College, Selcuk University, Konya, Turkey
[2] Faculty of Engineering and Architecture, Selcuk University, Konya, Turkey
[3] Department of Physical Education and Sport, Middle East
Technical University, Ankara, Turkey

1. INTRODUCTION

Developing skills improves performance in sport (Lees, 2002). Biomechanical analysis in soccer quantifies movement and helps skill development (Carling *et al.*, 2008). The penalty kick, a kind of free kick in soccer, is one of the most important and determining events of a game. The most commonly used technique is the instep kick (Shan and Westerhoff, 2005). Therefore, training to improve the skills and strategies of a free kick is essential for success. Previous studies (Bloomfield *et al.*, 1979; Asami and Nolte, 1983; Barfield, 1997) in young and mature professional players focused on the variables that predict the success of the instep kick. Elite soccer players develop a relatively constant movement strategy for the instep kick (Gainor *et al.*, 1978; Ben-Sira, 1980; Hay, 1996; Tsaousidis and Zatsiorsky, 1996; Orloff *et al.*, 2008). From the mechanical point of view, the instep kick is a multiplanar movement and the motion of a single segment within a linked system is nonlinear and cannot be solely attributed to muscle forces and moments acting on the lower extremity (Barfield *et al.*, 2002). Biomechanical analysis of a well executed instep kick requires the assessment of the whole lower extremity kinetic chain (Shan and Westerhoff, 2005; Orloff *et al.*, 2008).

Previous studies have focused on ball velocity as an indication of success in instep kicking (Barfield *et al.*, 2002; Smith *et al.*, 2006). However, kinematic analysis of this movement in relation to success has not been documented. The instep kick can be divided into three phases. These are the back swing, ball contact and the follow-through. Each phase should be executed effectively so that the ball is kicked to the desired location. Therefore, the aim of this study was to investigate the differences in ankle extension and knee flexion during penalty taking using the instep kick where success was indicated at four different locations, the upper left (UL), the lower left (LL), the upper right (UR), and the lower right (LR) inner corners of the goal. Understanding such differences will help soccer trainers to use this information to improve their players' instep technique.

2. METHODS

Five collegiate healthy male soccer players (mean ± SD: age, 21.5 ± 2.2 years; stature, 1.74 ± 0.06 m; body mass, 68.4 ± 4.6 kg; soccer experience, 10.4 ± 5.0 years) volunteered to participate in the investigation. Each player executed 5 successful instep penalty kicks to the targets (60x60 cm) placed at the 4 (UL, LL, UR, and LR) inner corners of the goal after 12 minutes' warm-up and stretching exercise. The mean of the five trials was used for statistical analysis. Five reflective markers were used to define the thigh, the shank and the foot segments. Markers were placed on the a) right anterior superior iliac spine, b) lateral condyle of the right knee, c) lateral malleolus, d) posterior of the calcaneus, and e) distal metatarsal head of the fifth phalanx. The instep penalty kicks were recorded at a stereoscopic view with two digital cameras (Dragonfly Express, Point Grey Research Ottawa, Canada) at a frame rate of 60 Hz. The cameras were placed at an approximately 90 degree angle to each other. A cage that covered the volume of 1.0 m x 1.0 m x 2.0 m at 12 control points was used to calibrate the space in which the instep penalty kicks were performed (Figure 1). Photogrammetric restitutions were conducted using the Pictran Software (Technet GmbH, Berlin, Germany). An adjustment process was provided in bundles of 6-8 control points. Three-dimensional coordinates of the marked points were defined after the adjustment process was completed. Ankle extension and knee flexion were determined from the images. The swing, ball contact and follow-through phases were assessed according to Barfield *et al.* (1998). One-way ANOVA and pairwise comparisons were conducted to determine the differences between ankle and knee strategies while kicking the ball to the four different locations in each kicking phase.

Figure 1. Marker locations during the instep penalty kick.

3. RESULTS

An ANOVA was conducted to determine the differences between ankle and knee kicking strategies during all three phases of the instep penalty kick. Results showed no significant difference in knee flexion when players kicked the ball to four inner corners of the goal in approach ($F_{3,16} = 0.11$; $P > 0.05$), ball contact ($F_{3,16} = 1.03$; $P > 0.05$) and follow-through ($F_{3,16} = 1.64$; $P > 0.05$) phases. Players tended to use the same knee flexion strategy for all targets.

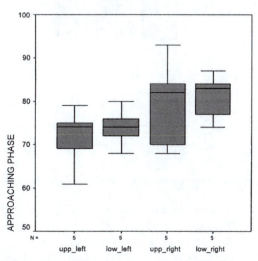

Figure 2. Knee strategy differences between UL, LL, UR, LR in approaching phase.

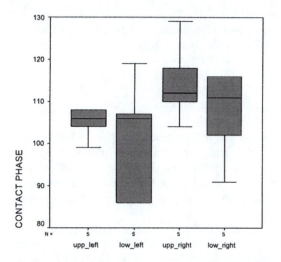

Figure 3. Knee strategy differences between UL, LL, UR, LR in the contact phase.

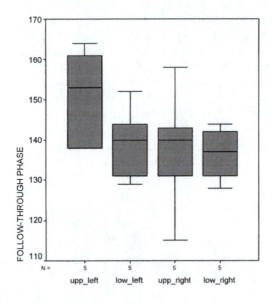

Figure 4. Knee strategy differences between UL, LL, UR, LR in the follow-through phase.

Results revealed no significant difference in ankle extension at the approaching phase (Figure 2). However, ankle extension differed significantly at ball contact ($F_{3,16}$ = 11.61; P < 0.001) and follow-through ($F_{3,16}$ = 3.07; P = 0.05) phases (Figures 3-4). Pairwise comparison of the contact phase revealed that players had significantly higher ankle extension when shooting the ball to the UR corner (77.6 ± 10.5°) and LR corner (84.4 ± 10.1°) when compared to the LL corner (56.2 ± 9.6°) and UL corner (59.6 ± 4.2°). Results for the follow-through phase revealed that players presented significantly higher ankle extension when shooting the ball to the LR corner (66.8 ± 7.1) compared to the LL corner (56.2 ± 6.0°) (Figures 5, 6 and 7).

4. DISCUSSION and CONCLUSION

Players presented a similar knee flexion strategy while shooting the ball to different locations in the goal in each phase of the instep kick. However, ankle extension at ball contact and follow-through phases was different when the LR and LL corners were targeted. To our knowledge, this is the first study relating lower extremity joint kinematics during an instep penalty kick to different locations in the goal.

The study was limited to collegiate male soccer players. The advantage of the study was the field application. Shinkai *et al.* (2007) focused on the motion of the foot at the contact phase of the kicking movement. They found that the foot was plantarflexed, abducted and everted during the contact with the ball. They also suggested that after foot contact with the ball it does not have any effect on the velocity of the ball (Shinkai *et al.*, 2007).

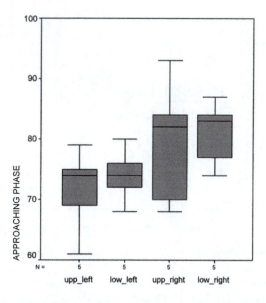

Figure 5. Ankle strategy differences between UL, LL, UR, LR in approaching phase.

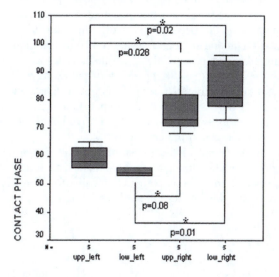

Figure 6. Ankle strategy differences between UL, LL, UR, LR in the contact phase.

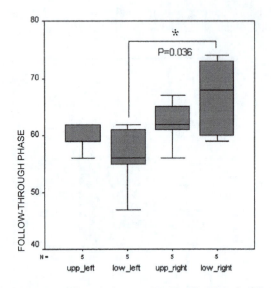

Figure 7. Ankle strategy differences between UL, LL, UR, LR in the follow-through phase.

Apriantono *et al.* (2006) investigated the effect of muscle fatigue on the kinematics of instep kicking and found that fatigue does have a negative effect on the kinematics of the movement. As the player develops fatigue in an actual game, the kinematics of his/her kicking action could be negatively affected and the probability of missing the penalty increases. Since the instep kick is the most commonly used technique in soccer, using a wrong technique may increase the energy consumption and cause fatigue in the muscles actively used for instep kicking which may also affect the precision of passes as well as a penalty kick. Well trained and highly skilled soccer players may use less energy by kicking the ball in an effective way with a well developed kinematic chain. Nunome *et al.* (2006) found that highly skilled soccer players have better inter-segmental motion for both the preferred and non-preferred leg.

In conclusion, our results indicate that players had similar knee flexion but not ankle extension kinematic strategies at the contact and follow-through phases of the instep penalty kick to different corners in the goal. The area of the goal targeted affects ankle extension kinematics but not knee flexion.

References

Apriantono, T., Nunome, H., Ikegami, Y. and Sano, S., 2006, The effect of muscle fatigue on instep kicking kinetics and kinematics in association football. *Journal of Sports Sciences*, **24**, pp. 951–960.

Asami, T. and Nolte, V., 1983, Analysis of powerful ball kicking. In *Biomechanics VIII*, edited by Matsui, H. and Kobayashi, K. (Champaign, Il: Human

Kinetics), pp. 695–700.

Barfield, W.R., 1997, Biomechanics of kicking. In *Textbook of Sports Medicine*, edited by Garrett, W.E. and Kirkendall, D.T. (Baltimore: Williams & Wilkins), pp. 86–94.

Barfield, B, 1998, The biomechanics of kicking in soccer. *Clinics in Sports Medicine*, **17**, pp. 711–728.

Barfield, W.R., Kirkendall, B.Y. and Yu, B., 2002, Kinematic instep kicking differences between elite female and male soccer players. *Journal of Sports Science and Medicine*, **1**, pp. 72–79.

Ben-Sira, D., 1980, A comparison of the mechanical characteristics of the instep kick between skilled soccer players and novices. Doctoral dissertation, University of Minnesota, Duluth, MN.

Bloomfield, J., Elliott, B. and Davies, C., 1979, Development of the soccer kick: A cinematographical analysis. *Journal of Human Movement Studies*, **3**, pp. 152–159.

Carling, C., Bloomfield, J., Nelsen, L. and Reilly, T., 2008, The role of motion analysis in elite soccer: contemporary performance measurement techniques and work rate data. *Sports Medicine*, **38**, pp. 839–862.

Gainor, B.J., Piotrowski, G., Puhl, J.J. and Allen, W.C., 1978, The kick: biomechanics and collision injury. *American Journal of Sports Medicine*, **6**, pp. 185–193.

Hay, J.G., 1996, *The Biomechanics of Sports Techniques*, 4th ed. (Englewood Cliffs, NJ: Prentice Hall).

Lees, A., 2002, Technique analysis in sports: a critical review, *Journal of Sports Sciences*, **20**, pp. 813–828.

Nunome, H., Ikegami, Y., Kozakai, R., Apriantono, T. and Sano, S., 2006, Segmental dynamics of soccer instep kicking with the preferred and non-preferred leg. *Journal of Sports Sciences*, **24**, pp. 529–541.

Orloff, H., Sumida, B., Chow, J., Habibi, L., Fujino, A. and Kramer, B., 2008, Ground reaction forces and kinematics of plant leg position during instep kicking in male and female collegiate soccer players. *Sports Biomechanics*, **7**, pp. 238–247.

Shan, G. and Westerhoff, P., 2005, Full-body kinematic characteristics of the maximal instep soccer kick by male soccer players and parameters related to kick quality. *Sports Biomechanics*, **4**, pp. 59-72.

Shinkai, H., Nunome H., Ikegami Y. and Isokawa M., 2007, Ball-foot interaction in impact phase of instep soccer kick. *Journal of Sports Science and Medicine*, Supplementum **10**, pp. 26-29.

Smith, C., Gilleard, W., Hammond, J. and Brooks, L., 2006, The application of an exploratory factor analysis to investigate the inter-relationships amongst joint movement during performance of a football skill. *Journal of Sports Science and Medicine*, **5**, pp. 517-524.

Tsaousidis, N. and Zatsiorsky, V., 1996, Two types of ball effect or interaction and their relative contribution to soccer kicking. *Human Movement Science*, **15**, pp. 861-876.

CHAPTER SEVEN

Biomechanics of the volley kick by the soccer goalkeeper

P.T. Gómez Píriz[1], M. Gutiérrez Dávila[2], D. Cabello Manrique[2] and A. Lees[3]

[1]Department of Physical Education, University of Sevilla, Sevilla, Spain
[2]Faculty of Sciences of Physical Activity and Sport,
University of Granada, Granada, Spain
[3]Research Institute for Sport and Exercise Sciences,
Liverpool John Moores University, Liverpool, UK

1. INTRODUCTION

The volley kick with self-release of the ball from the hands by the soccer goalkeeper is a frequent skill demonstrated in normal play. The goalkeeper kicks the ball aiming at one of his or her team-mates, and, given that it is initiated by the hands from the goal area, would be expected to have characteristics differing from that of a side foot or instep kick used during normal play. The analysis of this skill during the 1998 World Cup in France (Raya and Gómez Píriz, 2000) suggested a need to determine the biomechanical characteristics of this type of kick.

The experimental investigation of kicks made when the ball is stationary are numerous. These kicks include direct free-kicks with a defending wall (Wang and Griffin, 1997; López, 1995) and penalty shots (Fradua *et al.*, 1994; McMorris *et al.*, 1995; Levanon and Dapena, 1998). All of these research groups have used various kinematic variables to describe the kicks. Thus, kinematic analysis provides the means to characterise the biomechanics of kicking skills.

The goalkeeper's kick with a self-release of the ball from the hands can be identified as an open and sequential kinetic chain of segment rotations that should meet two objectives: a) project the ball at a certain speed and b) attain target accuracy. Therefore the aims of this study were to describe selected biomechanical variables, established through a three-dimensional analysis, of the soccer goalkeeper's volley kick with self-release of the ball from the hands in order to achieve targeting accuracy and establish any relationships between key variables.

2. METHODS

2.1 Participants

The incidental sample was made up of experienced elite soccer goalkeepers who were part of the playing staff of two Spanish first-division clubs (N = 9; height 1.84 ± 0.034 m; time as professional 6.6 ± 4.6 years and mass 82.48 ± 5.07 kg).

The observers who evaluated the degree of accuracy of each kick were two Superior Technicians in soccer (observers 1 and 2), former professional soccer players. The receiving player was an active professional player (10 years as a professional player).

2.2 Research variables

Photogrammetric and three-dimensional (3D) video techniques involving the filming and computerization of images were used. Accuracy was determined as described below. The most accurate kick from each player was taken for detailed biomechanical analyses. From these, several variables were determined. These were ball variables determined by the initial angle of projection and speed of the ball; temporal variables determined by key moments (T1-T4) for the kick as illustrated in Figure 1 and determined in Table 1; and performance variables including the length of the last step, thigh/trunk angle in the kicking leg at impact and hip and shoulder rotation angles about the vertical axis for each key moment.

Table 1. Definition of events and establishment of stages in the goalkeeper's kick with self-release of the ball from the hands.

Events	Definition	Stages			
T1	Moment in which the ball loses contact with the goalkeeper's hands	Stage A (Preparatory)			TIEMPO TIME (TT)
T2	Moment in which the kicking foot loses contact with the floor		Stage B (of fly)		
T3	Initial contact of the supporting foot with the ground			Stage C (of kick)	
T4	Initial contact of the kicking foot with the ball				

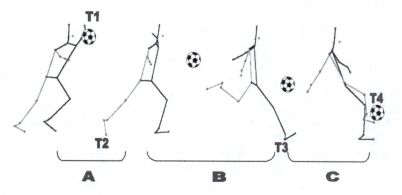

Figure 1. The soccer goalkeeper's volley kick with self-throw of the
ball from the hands (events and stages).

2.3 Data collection and analysis

Each trial had the following sequence: one player passes the ball with his hands so
the research participant (goalkeeper) can kick the ball towards a receiving field
player who moves at random to one of the three reception areas (Zones A, B and C
placed in the middle of the opponent's field and distributed equally along it) just
when the goalkeeper is adapting the ball with the hands. The accuracy of the kick
was determined on a five-point scale according to the criteria that the ball gets into
the correct reception zone and goes to its geometric centre, towards which the
receiver has moved simultaneously, with such a trajectory that it is not necessary to
control the ball in any way in order to move forward in the direction of the
opponent's goal area.

The goalkeeper kicked in a space (4x2x2 m) calibrated for 3D filming using
two video cameras (Sonics V-200 at 50 Hz). Selected trials were analysed using
the 3D software produced at the Biomechanical Laboratory of the University of
Granada (Gutiérrez and Soto, 1992). Data were smoothed with a fifth-degree
spline.

Official balls from the Spanish Soccer League were used and were in
compliance with Norm II of the soccer regulations and FIFA rules concerning
measures, weight and pressure and ball pressure gauge (Uhlsport Make).

2.4 Statistics

Descriptive statistics were used to provide means and standard deviations for
variables. Pearson's Product-Moment correlation was used to determine significant
correlations between variables at $P < 0.05$ (SPSS, version 15.0).

3. RESULTS and DISCUSSION

The descriptive data for the variables analysed are shown in Table 2. These show mean values for ball and temporal variables. The data for hip rotation and shoulder rotation angle are given at T2, T3 and T4. The thigh/trunk angle is given at T4 only. The relationship between the initial ball speed and accuracy is shown in Figure 2. The initial angle of projection of the ball (38.5°) was higher than when the ball is kicked from the ground in which a mean angle of projection of 18.9° and 6.1° for high and low trajectories respectively, has been reported (Prasas *et al.*, 1990). In the volley kick the foot is more able to get under the ball, enabling it to produce a higher angle of projection. The initial ball speed (31.8 m.s^{-1}) was also higher than that reported for kicking a stationary ball. Prasas and colleagues (1990) reported the initial speed of the ball was 21.6 m.s^{-1} in high-ball trajectories and 23.4 m.s^{-1} in low trajectories. Lees *et al.* (2004) reported 24.5 ± 1.39 m.s^{-1} for a maximal instep kick. The higher speed could be due to the greater amplitude of the movements in the kinetic chain, even when the participants were not asked for maximal speed kicks. A further explanation could be the better foot/ball contact position achieved when the foot is placed under the ball. Thus, the volley kick enables greater ball angles of projection and ball speeds to be reached, both of which would give the ball a greater range, one of the prime requirements of this type of kick.

The average step length was 1.53 m. Olson (1992) concluded that this variable, which in his studies ranged between 1.22 and 1.62 m, is related to leg length but was also linked to ball speed. In this research, no correlation was established between this variable and the initial speed of the ball. However, there was a correlation between hip rotation angle and step length (r=0.402, P=0.037); though low, this correlation seems logical. In any case, adopting a comfortable body position makes a better performance of the skill possible. Obtaining extreme levels in step length may lead to inappropriate positions when it comes to executing the kick accurately.

The thigh/trunk angle at T4 (moment of the contact with the ball) was 168°±4.6. Prasas and colleagues (1990) reported 174° and 150° in stationary-ball kicks for high and low ball trajectories, respectively. The angles for the high trajectories are similar to the angles obtained in the goalkeeper's volley kick with self-release of the ball from the hands.

The angular displacement of the hips at T4 (124°) was similar to that reported by Lees *et al.* (2004) who found angular displacement of the pelvis of 123° ± 34 in the maximum instep kick; these values are similar to the kick analysed in this study, although the research subjects were not asked to try for maximum speed.

The temporal values obtained confirm that we are dealing with a task that has a high execution time. In lengthy tasks (>1200 ms), the possibility of re-programming the response has been proved (Oña, 1994); this process occurs in the goalkeeper's volley kick with self-release of the ball from the hands. The total time of execution was 3177 ms, with stage C (see Figure 1) having an average duration of 1511 ms. Thus, even in the kick stage it would be possible to re-programme the motor response in accordance with the requirements of play. This has an influence on accuracy. While few of the kicks achieved the maximum score, the kicks analysed did show acceptable accuracy.

Table 2. Descriptive data of all the variables analysed in the soccer goalkeeper's volley kick.

	Starting angle of ball (°)	38.5° ± 3.3
	Initial speed (ms^{-1})	31.4 ± 2.6
	Step length (m)	1.53 ± 0.32
Execution time variable	Total time (ms)	3177 ± 277
	Stage A (preparatory, ms)	911
	Stage B (release, ms)	755
	Stage C (kick, ms)	1511
T4	Thigh/Trunk angle in T4	168.9° ± 4.6
	Hip rotation angle in T2	44.8° ± 23.9
	Hip rotation angle in T3	36.6° ± 17.5
	Hip rotation angle in T4	124.1° ± 20.5
	Shoulders rotation angle in T2	68.8 ± 10.6
	Shoulders rotation angle in T3	57.2 ± 19.1
	Shoulders rotation angle in T4	105.7 ± 11.7

4. CONCLUSIONS

It is concluded that while several of the biomechanical characteristics in the volley with ball release from the hands are similar to kicking a stationary ball, there is greater time available to perform this type of kick which may influence the accuracy achievable. The impact position of the foot on the ball which is thought to provide a better foot/ball contact is a means whereby a greater projection angle and ball speed can be achieved, and through these a greater range.

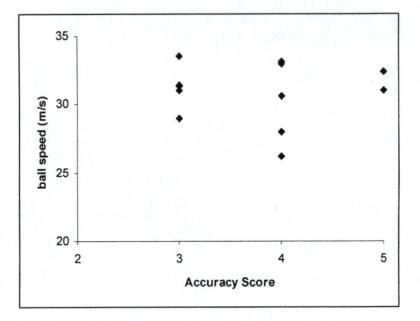

Figure 2. Relationship between the initial speed of the ball and accuracy variables.

References

Fradua, L., Raya, A., Pino, J. and Arteaga, M., 1994, Improving the goalkeeper's performance in penalty situations. *Science and Football*, **8**, pp. 25-27.

Gutiérrez, M. and Soto, V., 1992, Biomechanical analysis of the kinetic chain involved in the instep kick in soccer. *Archivos de Medicina del Deporte*, **34**, pp. 165-171.

Lees, A., Kershaw, L. and F. Moura, 2004, The three-dimensional nature of the maximal instep kick in soccer 'Part I: Biomechanics'. *Journal of Sports Sciences*, **22**, pp. 485 – 500.

Levanon, J. and Dapena, J., 1998, Comparison of the kinematics of the full-instep and pass kicks in soccer. *Medicine and Science in Sports and Exercise*, **30**, pp. 917-927.

López, M., 1995, *Biomechanical analysis of direct free kicks with wall in soccer.* Thesis on microfiche, Facultad de Ciencias de la Actividad Física y el Deporte, University of Granada.

McMorris, T., Hauxwell, B. and Holder, T., 1995, Anticipation of soccer goalkeepers when facing penalty kicks to the right and left of the goal using different kicking techniques. *Applied Research in Coaching and Athletics Annual*, pp. 32-43.

Olson, K. R., 1992, Leg length and distance of approach steps for the optimal soccer style placekick, publication on microfiche: University of Oregon.

Oña, A., 1994, *Motor Behaviour: Psychological Bases of Human Movement* (Granada, Spain: University of Granada).

Prasas, S.G., Teradus, J. and Nathau, T.A., 1990, Three-dimensional kinematic analysis of high and low trajectory kicks in soccer. *VIII Symposium ISSB*, pp. 145-149.

Raya, A. and Gómez Píriz, P.T., 2000, Analysis of the instep kick in the soccer goalkeeper. *Training Fútbol*, **53**, pp. 16-31.

Wang, J. and Griffin, M., 1997, Kinematic analysis of the soccer curve ball shot. *Strength and Conditioning*, **19**, pp. 54-58.

CHAPTER EIGHT

Change of ball impact technique in instep kicking with physical growth

H. Shinkai[1], H. Nunome[2], H. Suzuki[1], H. Suito[1], N. Tsujimoto[1],
T. Kumagai[1] and Y. Ikegami[2]

[1]Graduate School of Education and Human Development,
Nagoya University, Japan
[2] Research Centre of Health, Physical Fitness and Sports,
Nagoya University, Japan

1. INTRODUCTION

Ball impact technique is an important factor in kicking the ball to achieve a fast velocity. The duration of ball impact during instep kicking in soccer lasts approximately for 10 ms and rapid deformations of the foot occur during this time (Asai *et al.*, 2002; Nunome *et al.*, 2006; Shinkai *et al.*, 2008). Obviously, lower sampling rates (100 to 500 Hz) used in the previous studies (Asami and Nolte, 1983; Rodano and Tavana, 1993) are insufficient to describe the movement characteristics in this phase adequately. To date, only a few studies have used a high enough sampling rate (1000 to 5000 Hz) in order to describe the foot and ball motion during ball contact for instep kicking (Asai *et al.*, 2002; Nunome *et al.*, 2006; Shinkai *et al.*, 2008).

Ball impact is a process for transferring the momentum of the kicking limb to the ball. Through daily training, young footballers learn to impart the foot and leg momentum into the ball effectively, in which ankle rigidity of the kicking limb has been considered a vital factor (Asami and Nolte, 1983; Rodano and Tavana, 1993; Lees and Nolan, 1998). Several attempts have been made to describe the process of how players achieve skilful kicking technique. With regard to the swing of the kicking leg, it has been shown that the maximum angular velocity of the shank and the values at ball contact increase with age (Bloomfield *et al.*, 1979; Luhtanen, 1988). However, the foot and ankle motion during ball impact and those changes from childhood to adolescence in footballers have never been systematically investigated. As physical maturity is quite significant during this time, it can be hypothesized that some developmental factors affect the change of ball impact characteristics.

The purpose of the present study, therefore, was to investigate the change of ball impact characteristics in instep soccer kicking with physical development of young players using ultra-high-speed sampling procedures.

2. METHODS

Twenty-one young male footballers categorized in seven age groups from 9 to 15 years old (three persons for each group) participated in the present study. All players belonged to the youth academy of a Japanese professional soccer club. The body mass of the players increased with age (see Table 1). Ethics permission was obtained from the Human Research Committee of the Research Centre of Health, Physical Fitness and Sports of Nagoya University. We also obtained agreement to participate from subjects, their parents and their coaches.

After an adequate warm-up, subjects were instructed to perform three consecutive maximal instep kicks using their dominant (right) leg with a free approach run on an artificial turf pitch. The target (0.88 m square) was set 5 m ahead and just above ground level. One successful shot that hit the target and had a good foot/ball impact was selected for each subject for analysis.

A FIFA approved soccer ball of size four (mass = 0.374 kg, inflation = 700 g/cm^2) or size five (mass = 0.42 kg, inflation = 900 g/cm^2) was used for subjects between 9 and 12 years old and between 13 and 15 years old, respectively. All subjects wore the same type of soccer shoes for outdoor practice to minimize the effects of type of shoe on the interaction between foot and ball during ball contact. Two electrically synchronized ultra-high-speed video cameras (Photron Ltd., FASTCAM-512 PCI) were set up on the diagonally kicking leg (right) side and diagonally backward. The sampling rate of the cameras was set at 2000 Hz to capture the foot and ball motion during ball contact. White markers were securely fixed onto the ball and over several anatomical landmarks on the lateral side of the kicking limb: head of fibula, lateral malleolus, lateral side of calcaneus, fifth metatarsal base and fifth metatarsal head.

The direct linear transformation (DLT) method was used to obtain the three-dimensional coordinates of each marker. Tri-axial angular motion of the foot (plantar/dorsal flexion, abduction/adduction, inversion/eversion) was calculated using the following vector operation. The segment vector of the shank (S_{Shank}) pointing from lateral malleolus towards the head of fibula and the segment vector of the foot (S_{Foot}) pointing from the lateral side of calcaneus towards the fifth metatarsal head were defined. The plantar-dorsal flexion angle was defined as the angle between S_{Shank} and S_{Foot} on the plane perpendicular to the vector (V_{FS}) made by the vector product of S_{Shank} and S_{Foot}. The abduction-adduction angle was defined as the angle between the Y axis (back and forth direction) and the vector (V_F) defined by the vector product of the vector pointing from the lateral side of calcaneus towards the lateral malleolus and S_{Foot} on the plane perpendicular to S_{Shank}. The inversion-eversion angle was defined as the angle between S_{Shank} and V_F on the plane perpendicular to the vector defined by the vector product of S_{Shank} and V_{FS}. Angular displacements during ball contact, whose criterion value was the angle of the foot at the instant of the initial ball contact, were also calculated. All angle data were digitally smoothed by a fourth-order Butterworth low-pass filter at 200 Hz.

The contact time between the foot and the ball was measured from the lateral side video image by counting the number of frames. The foot and ball velocities were represented by the velocity of the fifth metatarsal base and the centre of the ball, respectively. Both velocities were computed for all components (XYZ) as the slope of the linear regression lines fitted to their non-filtered displacements (10 points just before and after ball contact) from which the absolute values were computed. The ball/foot velocity ratio was represented as the index of efficiency of momentum transfer from the kicking limb to the ball. Moreover, to quantify how much the weight was used for ball impact, the effective striking mass of the kicking limb was estimated using the equation of conservation of momentum (Plagenhoef, 1971). For comparison with this value, the mass of the player's foot was obtained from body segment parameters of Japanese children (Yokoi *et al.*, 1986).

Table 1. Anthropometric characteristics of the subjects in different age groups
(e.g. U-9 means the age of 9 and under).

Age group (years)	U-9	U-10	U-11	U-12	U-13	U-14	U-15
Number of subjects	3	3	3	3	3	3	3
Body mass (kg)	24.7 (2.8)	29.6 (0.4)	32.2 (1.2)	40.0 (1.6)	47.2 (3.8)	51.6 (2.0)	63.3 (6.5)
Shoe size (cm)	20.7 (1.2)	21.2 (0.3)	22.5 (0.9)	23.8 (1.4)	26.2 (0.3)	26.2 (0.3)	27.0 (0.9)

Values are mean (SD)

3. RESULTS

The average contact time between foot and ball was 9.0 ± 0.5 ms (range from 8.5 to 10.0 ms) and was consistent for all age groups. Foot velocity before ball impact (range from 14.6 to 20.5 $m.s^{-1}$), the resultant ball velocity (range from 14.5 to 28.9 $m.s^{-1}$) and ball/foot velocity ratio (range from 0.98 to 1.42) increased with age, whereas the reduction rate of foot velocity between before and after ball impact (range from 27.9 to 53.5 %) decreased with age. A strong positive correlation ($r = 0.90$, $P < 0.01$) was found between the foot velocity before ball impact and the body mass of the players. The resultant ball velocity was strongly correlated with the foot velocity before ball impact ($r = 0.94$, $P < 0.01$) and the body mass of the players ($r = 0.94$, $P < 0.01$).

During ball contact, the kicking foot was forced into plantar flexion (6.9 ± 4.9 deg), abduction (7.1 ± 4.5 deg) and eversion (2.6 ± 2.8 deg) and these motions were commonly observed in most of the trials regardless of age. Figure 1 shows the relationship between angular displacement of foot-plantar flexion during ball contact and ball/foot velocity ratio. As shown, there was no significant relationship between these variables. Furthermore, there was also no significant relationship between the angular displacements for the other directions (foot abduction and eversion) and the ball/foot velocity ratio. Likewise, angular displacements for all directions were not correlated with the resultant ball velocity.

The average effective striking mass of the kicking limb was 1.28 ± 0.38 kg (range from 0.69 to 2.14 kg). This variable was strongly correlated with the body mass of the players ($r = 0.96$, $P < 0.01$) and ball/foot velocity ratio ($r = 0.91$, $P < 0.01$) (see Figure 2). The average foot mass of players estimated from Yokoi and colleagues' study was 0.77 ± 0.26 kg (range from 0.41 to 1.34 kg) and the mass of the shoe was 0.27 ± 0.04 kg (range from 0.20 kg with smallest size (20 cm) to 0.33 kg with largest size (27.5 cm)). The average mass of the shod foot was 1.04 ± 0.30 kg (range from 0.62 to 1.67 kg) and it corresponded to 81.3 ± 7.5 % of the effective striking mass.

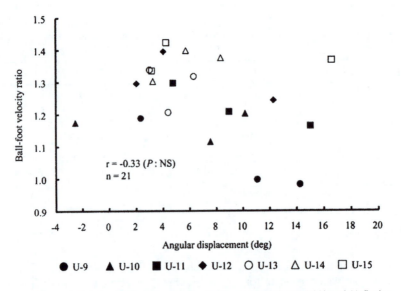

Figure 1. Relationship between angular displacement of plantar (+)/dorsal (-) flexion during ball contact and ball-foot velocity ratio.

4. DISCUSSION

To date, there are only a few studies focused on the relationship between kicking performance and physical development. Luhtanen (1988) published the only study that reported the ball velocity of instep kicking of skilled young footballers from 9 to 18 years old. Thus, the results of his study can be used for a comparison with those of the present study. The resultant ball velocity of the present study showed distinctively higher values than those reported by Luhtanen (1988). In his study, the ball velocity of the youngest group (9 to 11 years old) was markedly slower than the average resultant ball velocity of 9 to 11 years old in the present study (14.9 ± 1.7 m.s^{-1} vs. 18.4 ± 2.5 m.s^{-1}). Moreover, the maximum ball/foot velocity ratio (1.42) of the present study was comparable to that of Nunome *et al.* (2006) (approximately 1.35) and Shinkai *et al.* (2008) (1.40 ± 0.07) who examined skilled adult footballers. These results suggest that the subjects of the present study have a skilful kicking technique.

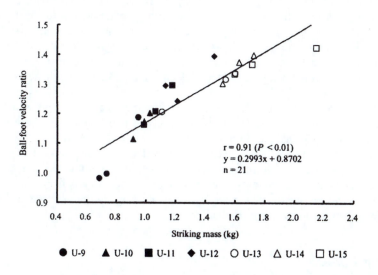

Figure 2. Relationship between effective striking mass of the kicking limb and ball-foot velocity ratio.

Shinkai *et al.* (2008) reported three-dimensional foot motion during ball contact for mature adult footballers. The results of the present study confirmed similar aspects of the foot motion for younger footballers undergoing the physical growth process. Shinkai *et al.* (2008) also reported that the magnitude of the ball reaction force acting on the kicking foot reached nearly 3000 N which was comparable to four times as much as the player's body mass. It can be assumed that this magnitude of force was so large that the muscles around the ankle cannot control the ankle motion. Thus, it can be concluded that the foot angular motions observed during ball contact were most likely passive motions. It has been suggested by several studies (Asami and Nolte, 1983; Rodano and Tavana, 1993; Lees and Nolan, 1998) that the state of rigidity of the foot is a vital factor for good foot/ball impact. Asami and Nolte (1983) showed that the state of rigidity of the forefoot during ball impact influences the resultant ball velocity. In contrast to this, the present study showed that the foot angular displacements for all directions have no relation to the ball/foot velocity ratio and the resultant ball velocity. These discrepant results between the two studies may be explained by the difference in data selection criteria. Although only successful shots with a good foot/ball impact were selected for analysis in the present study, the criteria were unexplained in the study of Asami and Nolte (1983), in which some trials with a large foot deformation were included for analysis.

In the present study, a positive strong correlation (r = 0.94, P < 0.01) between the player's body mass and resultant ball velocity was found. This result was consistent with the report of Luhtanen (1988), in which the release ball velocity correlated significantly with the body weight of the players (r = 0.93). The results of these two studies indicate that the body mass of players greatly influences the resultant ball velocity. Moreover, the effective striking mass which represents how much weight is used for ball impact, was strongly correlated with the ball/foot velocity ratio (r = 0.91, $P < 0.01$). As the mass of the shod foot corresponded to 81.3 ± 7.5 % of the effective striking mass, it can be suggested that if the foot hit the ball with an adequate foot position (i.e. approximate location of its centre of mass) then the ball impact of the instep kick can be regarded as a collision between shod foot and ball.

To enhance the striking mass, increasing the mass of the foot is essential. It seems that the mass of the kicking foot (as a proportion of the body mass) increased with physical development and was strongly correlated with the efficiency of ball impact. Thus, it can be suggested that physical size of players greatly affects the quality of ball impact technique, in which younger players with a lighter foot have a disadvantage in producing a good foot/ball impact. Soccer coaches should be aware of this fact when evaluating the ball impact technique especially for growing younger footballers.

5. CONCLUSION

In the present study, the change of the ball impact characteristics with physical development was systematically investigated using a filming procedure that was adequate for the purpose. The kicking performance of skilled young footballers increased with physical development and the efficiency of ball impact was correlated strongly with the effective striking mass of the kicking limb. As the mass of the shod foot was approximately equivalent to the effective striking mass, it is suggested that the weight of the foot has a substantial influence on the efficiency of ball impact.

References

Asai, T., Carré, M. J., Akatsuka, T. and Haake, S. J., 2002, The curve kick of a football I: impact with the foot. *Sports Engineering*, **5**, pp. 183-192.

Asami, T. and Nolte, V., 1983, Analysis of powerful ball kicking. In *Biomechanics VIII-B*, edited by Matsui, H. and Kobayashi, K. (Champaign, Illinois: Human Kinetics Publishers), pp. 695-700.

Bloomfield, J., Elliott, B. C. and Davies, C. M., 1979, Development of the soccer kick: A cinematographical analysis. *Journal of Human Movement Studies*, **5**, pp. 152-159.

Lees, A. and Nolan, L., 1998, The biomechanics of soccer: A review. *Journal of Sports Sciences*, **16**, pp. 211-234.

Luhtanen, P., 1988, Kinematics and kinetics of maximal instep kicking in junior soccer players. In *Science and Football*, edited by Reilly, T., Lees, A.,

Davids, K. and Murphy, W.J. (London: E and FN Spon), pp. 441-448.

Nunome, H., Lake, M., Georgakis, A. and Stergioulas, L.K., 2006, Impact phase kinematics of instep kicking in soccer. *Journal of Sports Sciences*, **24**, pp. 11-22.

Plagenhoef, S., 1971, *Patterns of Human Motion: A Cinematographic Analysis* (Englewood Cliffs, New Jersey: Prentice-Hall).

Rodano, R. and Tavana, R., 1993, Three-dimensional analysis of instep kick in professional soccer players. In *Science and Football II*, edited by Reilly, T., Clarys, J. and Stibbe, A. (London: E and FN Spon), pp. 357-363.

Shinkai, H., Nunome, H., Ikegami, Y. and Isokawa, M., 2008, Ball-foot interaction in impact phase of instep soccer kicking. In *Science and Football VI*, edited by Reilly, T. and Korkusuz, F. (Abingdon: Routledge), pp. 41-46.

Yokoi, T., Shibukawa, K. and Ae, M., 1986, Body segment parameters of Japanese children. *Japan Journal of Physical Education, Health and Sport Sciences*, **31**, pp. 53-66.

CHAPTER NINE

Kinetic similarity in side-foot soccer kicking with the preferred and non-preferred leg

R. Kawamoto, O. Miyagi and J. Ohashi

Department of Sports Science, Daito Bunka University, Saitama, Japan

1. INTRODUCTION

Developing kicking skill with the non-preferred leg is an important task for soccer players. To understand the biomechanical differences between the preferred and non-preferred leg during soccer kicking, some studies have been undertaken (Capranica *et al.*, 1992; Dörge *et al.*, 2002; Nunome *et al.*, 2006). However, kinetic similarity during side-foot kicking between the preferred and non-preferred leg has not been quantitatively investigated. The aim of this study was to investigate whether the kinetic similarity during soccer side-foot kicking between the preferred and non-preferred leg is dependent on soccer experience.

2. METHODS

Seven experienced male soccer players and eight inexperienced players participated in the study. The experienced players had played for 10 to 15 years. None of the inexperienced participants had received special training in soccer.

Eight optoelectronic cameras of a motion capture system (Motion Analysis Corp., California, USA) were placed within the laboratory. Prior to data acquisition, 24 reflective markers were attached to the anatomical landmarks of each participant's body. Two reflective stickers were also placed on the side of the ball to determine the timing of impact and the resultant ball velocity after the impact.

All kinematic data were acquired by the cameras at 200 Hz. The participants were instructed to perform side-foot kicks along the ground with maximum effort with an eye on the target line 8.0 m away from the initial ball position. They were also instructed to kick the ball with a single-step run-up as a preparation for kicking. All participants had five trials for each leg. The trial in which the ball speed was the fastest was selected for analysis.

Captured motion data were reconstructed into three-dimensional coordinates by the direct linear transformation method. A relatively high frequency (20 Hz) was selected as the cut-off frequency of the Butterworth-type digital low-pass filter. This procedure was to minimize any distortion of the kinematic data around impact.

The kicking leg was modeled as a link-segment model composed of the foot, shank and thigh. Using this model, the ankle, knee, and hip joint torques were calculated through inverse dynamics (Winter, 2005). Dorsi/plantar flexion moment at the ankle, flexion/extension moment at the knee, and flexion/extension, adduction/abduction, and internal/external rotation moments at the hip were computed in this study.

For each subject, correlation analyses were performed between the preferred and non-preferred leg using the normalized time-series data of the joint torques during kicking (50 samples from the maximum hip extension to ball impact). The correlation coefficient shows the degree of similarity of the joint torque between the preferred and non-preferred leg. Cross-correlation analysis is suitable to determine the similarity of two time-series curves. The correlation analysis is acceptable because the time was normalized and synchronized before the analysis. The mean correlation coefficient for each joint moment was statistically compared between the experienced and inexperienced group using an unpaired t-test. The kicking foot velocity and the ball velocity were also compared between the groups and between the legs. A two-way ANOVA was used to test these differences. Statistical significance was set at $P < 0.05$.

3. RESULTS and DISCUSSION

Mean kicking foot velocity and mean ball velocity were significantly higher in the experienced group than in the inexperienced group ($P < 0.001$, Table 1). These results confirm that the side-foot kicking performance of the experienced players was superior to that of the inexperienced players. In contrast, these velocities were not significantly different between the preferred and non-preferred leg regardless of soccer experience.

Table 1. Kicking foot velocity, ball velocity, and ratio of them.

	Experienced		Inexperienced	
	Preferred	Non-preferred	Preferred	Non-preferred
Foot velocity (m.s^{-1})	12.9 ± 0.6	12.6 ± 0.6	10.7 ± 0.7	9.7 ± 1.0
Ball velocity (m.s^{-1})	22.3 ± 1.8	23.4 ± 1.7	14.5 ± 2.1	13.1 ± 4.2
Ratio	1.73 ± 0.12	1.85 ± 0.12	1.36 ± 0.19	1.33 ± 0.32

Figure 1 shows stick figures for typical subjects. Regardless of soccer experience, side-foot kicking motions were generally similar between the preferred and non-preferred leg. Figure 2 shows mean time-series curves of the joint torques for the experienced group. Figure 3 shows the same torque curves for the inexperienced group. All torque curves matched well between the preferred and non-preferred leg for both groups. Table 2 shows mean correlation coefficients representing the similarity of the joint torques between the legs. In the experienced group, a very high positive correlation was found for all joint torques (r >= 0.88, P < 0.001). In the inexperienced group, high positive correlations were also found although the correlation of the ankle dorsi/plantar flexion torque (r = 0.67, P < 0.01) was lower than for the other joint torques (r >= 0.78, P < 0.001).

For any joint torques, mean correlation coefficients were not significantly different between the experienced and inexperienced group.

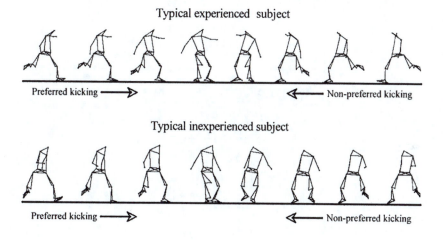

Figure 1. Stick figures for typical subjects.

The joint torque curves matched well for the two legs not only in the experienced group but also in the inexperienced group. This result suggests that coordination of the side-foot kicking motion is similar between legs regardless of practice. The coordination may depend on individual characteristics such as body shape, flexibility, and skill of the kicking motion. To develop side-foot kicking skill effectively, it could be important not to interfere with these individual characteristics.

The torque exertion of hip flexion was considerably different between the experienced and inexperienced groups, although the similarity of this torque was not significantly different (Figure 2, Figure 3 and Table 2). This result suggests that the experienced soccer players had acquired the skill of strong side-foot kicking using sufficient torque of hip flexion. In practical situations in soccer, the frequency of kicking is not the same for the two legs because most players would kick a ball with the preferred leg. Nevertheless, the joint torque curves matched well for the two legs. This result implies that bilateral transfer of the kicking motion may also occur through the specialized practice of soccer.

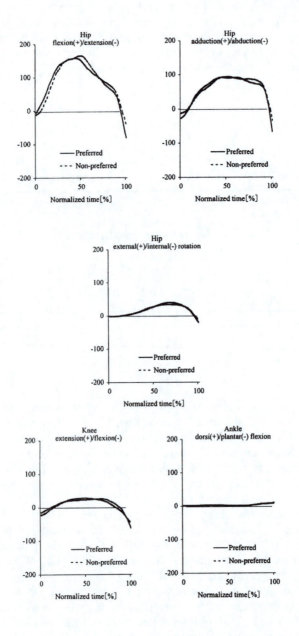

Figure 2. Mean time-series curves of joint torques (Nm) for the experienced group.

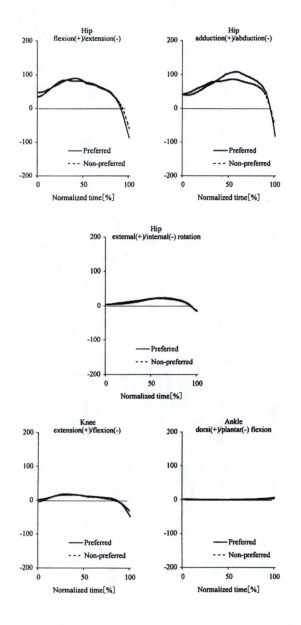

Figure 3. Mean time-series curves of joint torques (Nm) for the inexperienced group.

Table 2. Mean correlation coefficients for both groups.

Joint torque	Experienced (n=7)	Inexperienced (n=8)
Ankle dorsi/plantar flexion	0.95 ± 0.07	0.67 ± 0.44 [n.s.]
Knee extension/flexion	0.88 ± 0.11	0.79 ± 0.15 [n.s.]
Hip extension/flexion	0.93 ± 0.03	0.84 ± 0.13 [n.s.]
Hip adduction/abduction	0.88 ± 0.11	0.86 ± 0.09 [n.s.]
Hip internal/external rotation	0.95 ± 0.06	0.91 ± 0.04 [n.s.]

Mean ± S.D.

Acknowledgements

This research was partially supported by the Ministry of Education, Science, Sports and Culture, Grant-in-Aid for Young Scientists (A), 19680029, 2008.

References

Capranica, L., Cama, G., Fanton, F., Tessitore, A. and Figura, F., 1992, Force and power of preferred and non-preferred leg in young soccer players. *Journal of Sports Medicine and Physical Fitness*, **32**, pp. 358-363.

Dörge, H.C., Anderson, T.B., Sorensen, H. and Simonsen, E.B., 2002, Biomechanical differences in soccer kicking with the preferred and the non-preferred leg. *Journal of Sports Sciences*, **20**, pp. 293-299.

Nunome, H., Ikegami, Y., Kozakai, R., Apriantono, T. and Sano, S., 2006, Segmental dynamics of soccer instep kicking with preferred and non-preferred leg. *Journal of Sports Sciences*, **24**, pp. 529-541.

Winter, D.A., 2005, *Biomechanics and Motor Control of Human Movement* (3rd ed.), (New York: John Wiley and Sons).

CHAPTER TEN

A new valid shock absorbency test for artificial turf

H. Nunome[1], Y. Ikegami[1], T. Nishikawa[2] and T. Horio[2]

[1]Research Centre of Health, Physical Fitness and Sports,
Nagoya University, Nagoya, Japan
[2]SRI Hybrid Inc., Kobe, Japan

1. INTRODUCTION

The use of artificial turf is common for many sports, in particular association football (soccer). It seems that third generation artificial turf (3-g turf) possesses a more "natural turf"-like appearance and properties such as realistic ball-surface and shoe-surface interaction. Football organizations FIFA and UEFA recently accepted the use of 3-g turf for official and international tournaments. Moreover, the possible use of the 3-g turf for the next World Cup tournament is currently under consideration. However, there has been a concern that some of the mechanical characteristics of artificial sports surfaces may be linked to acute or chronic sports injuries. Several authors (Nigg, 1983; 1990) have already suggested that mechanical properties of artificial sports surfaces may be associated with sports injuries.

Various factors are important in the selection of a surface. Relevant surface properties include cushioning ability, friction characteristics and the influence on energy loss (Dixon *et al.*, 1998). Of these aspects, the cushioning ability can be considered as a key feature to prevent injuries. From the FIFA guidelines (2005), "Shock absorbency" of the 3-g turf has been assessed by a simple mechanical test (DIN test). However, it has been suggested by several authors (Blackburn *et al.*, 1995; Dixon, 1998; Nigg, 1990) that the current, conventional test procedure does not reflect the actual loading action occurring in sports movements. The present study aims to re-examine how much the current mechanical test procedure is valid to evaluate the shock absorbency of the 3-g turf and to establish a new test procedure which precisely reflects the acute load imposed by human sports actions.

2. METHODS

The present study was composed of a series of three experiments which consisted of a human experiment, re-examination of the standard test and finally the development of a novel test procedure. These tests required the construction of 3-g turf trays (90×90 cm) with different infill components and depths (Table 1).

Table 1. Details of tested 3-g turf trays.

Tray	Infill component	Infill thickness
Regular	Four layers: sand; rubber tip; sand; rubber tip	40 mm
Rubber 40	Rubber tip	40 mm
Rubber 15	Rubber tip	15 mm
Sand 40	Sand	40 mm

2.1 Human experiment

To obtain the baseline of the acute loads acting on the body, four healthy male subjects (height: 1.743 ± 0.063 m; mass: 71.6 ± 3.5 kg; age: 28.8 ± 7.8 years) volunteered for the study. A "hard landing" was chosen as a task which represents an acute human sports action. Because footballers are often forced to land from a jumping header with awkward postures, this sports action produces the biggest impact both on their body structures and the playing surface. Informed written consent was obtained from each subject.

All subjects wore the same type of training shoe designed for artificial turf (Del mundo, Puma Inc.) and performed a landing from a 50-cm height onto the bare rigid surface of the force platform three times. They were instructed to land with minimal shock attenuation. Ground reaction force was measured using a Kistler force platform sampling at 1000 Hz.

2.2 Re-examination of the standard test

Validity of the standard DIN 18032 test was re-examined. In this test, a 20 kg mass is released from a 55-mm height onto a load cell with spring (spring constant = 2000 ± 60 N.mm^{-1}) on the top (Figure 1). Using the FIFA guidelines (2005), the test was conducted on the 3-g turfs with different infill components. The test was also conducted on a bare rigid concrete surface to obtain the baseline of shock absorbency (corresponding to 0% of shock absorbency).

2.3 Development of new test procedure

For reproducing the load similar to the human hard landing, a test rig was newly developed (Figure 2). To produce a constant vertical load using the simplest structure, the rig applied a drop mass system using a length-adjustable pendulum.

Figure 1. Standard DIN test.

Figure 2. New test rig.

A solid tyre was used for the top instead of the spring system applied for the standard test. This type of tyre was chosen over air tube tyres because it was expected to be unaffected by the size of the contact area due to different loads.

Several types of the 3-g turf tray with different infill components and depths (Table 1) were tested in two loading conditions (approximately 7000 and 11000 N to the bare rigid surface of the force platform). The high loading condition resembles the load of a human hard landing and the low loading condition replicates the load of the standard test when it was conducted onto a concrete surface.

3. RESULTS

Figure 3 shows a typical example of the loading aspect from a human hard landing. The average peak magnitude was 11484 $\pm$1711 N (ranging from 9406 to 13524 N). As shown, one clear force peak was observed during the hard landing.

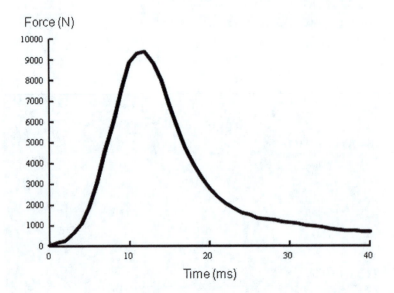

Figure 3. Typical loading curve during a human hard landing.

Figure 4 shows typical examples of the loading aspect of the standard test for concrete and different types of the 3-g turf. As shown, the initial loading rates were overlapping for all conditions and except for the concrete and sand infill tray, multiple peaks were produced during loading. The first peak observed for the three types of the 3-g turfs (Rubber 40, Rubber 15 and Regular) occurred at the same time and with the same magnitude. Moreover, the peak load of the standard test was distinctly smaller than that caused by the human landing.

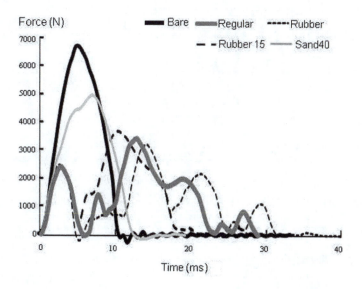

Figure 4. Typical loading curves during standard DIN test.

Figure 5 shows the average (±SD) loading curves generated by the new test rig in high loading (top) and low loading (bottom) conditions. These two loading conditions approximately simulated the peak load of the human hard landing and that of the standard test, respectively. As shown, one clear peak resembling human landing was observed for all types of 3-g turf whereas the standard test created multiple peaks for three of the four types of 3-g turf. In the high loading condition, the 3-g turfs attenuated the peak impact load to 55% and extended the time to peak up to 300% compared with the bare rigid surface.

The difference in shock attenuation ability (peak and loading rate) between the 3-g turfs was clearly illustrated in the high loading condition whereas the difference was unclear for some of the 3-g turfs (Rubber 40, Rubber 15 and Regular) in the lower loading condition which reproduced a similar load to the standard test.

4. DISCUSSION

In the present study, the validity of the standard DIN test, which has been accepted as the standard test procedure, to measure "shock absorbency" for the 3-g turf was re-examined in detail. An attempt was also made to develop a new test procedure which closely reproduced the loads acting on the body during landing from a jump.

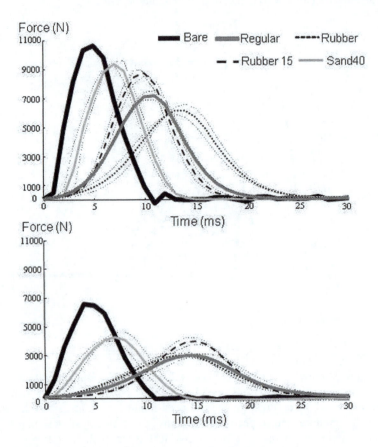

Figure 5. The average (±SD) change of the load applied by the new test rig in high
loading (top) and low loading (bottom) conditions.

4.1 Validity of the standard test

As shown in Figure 4, the standard test produced multiple peaks for three types of
the 3-g turf during loading. Of these, the peak magnitude (typically observed as the
second peak) has been evaluated as the only parameter to represent shock
absorbency. This unique loading aspect indicates that the force measuring part (test
foot with the load cell) oscillates between the impacting mass and the surface of
the 3-g turf during loading. Nigg (1990) examined the validity of the standard test,
and reported that accelerative motion of the test foot corresponds to a possible
inertia term up to 540 N which must be taken into account when assessing
surfaces. Moreover, the initial loading rates were overlapping for all changes

(including the change for the concrete surface) and the first peak observed for the three different infill 3-g turfs (Rubber 40, Rubber 15 and Regular) occurred at the exact same time and with the same magnitude (Figure 4). It is obvious that these unique initial loading aspects do not reflect the shock attenuating property of the 3-g turfs but reflects some mechanical property internal to the measurement system. As Junqua *et al.* (1983) illustrated, the spring system is most likely responsible for these unique loading aspects thereby possibly obscuring the representative shock absorbency properties of the 3-g turfs. Likewise, the magnitude of the load caused by the standard test (approximately 7000 N on the concrete and 3500 N on the 3-g turf) was apparently smaller than that of the human hard landing (Figure 3). It can be assumed that use of the standard DIN test may be inappropriate to evaluate the shock attenuation of the 3-g turf.

4.2 New test procedure

In contrast, the newly developed test rig succeeded in reproducing a similar single force peak loading curve to a human hard landing with reasonable magnitudes (Figure 5). The new test rig allows the evaluation of not only the peak magnitude but also the loading rate which is assumed to be an important measure of human subjective rating of impact severity. The high loading condition (approximately 11000 N) allowed the differentiation of shock attenuation properties (peak and loading rate) between the 3-g turfs with different infill components whereas the difference was unclear for some of the 3-g turfs in the lower loading condition (approximately 7000 N). This finding demonstrated that the high loading condition should be taken into account to evaluate the shock attenuation of 3-g turf.

5. CONCLUSION

The standard DIN test appeared to be inappropriate for evaluating the shock absorbency of several types of 3-g turf. For high loads such as those experienced during jump landings, the new test procedure demonstrated its distinct advantage in evaluating the shock absorbency of the 3-g turf.

Acknowledgements

A part of this study was financially supported by Grant-in-Aid for Exploratory Research of Japan Society for the Promotion of Science (No: 19650170) and SRI Hybrid Inc. Japan.

References

Blackburn, S., Nicol, A. C. and Walker, C., 2005, Development of a biomechanically validated turf testing rig. In *Proceedings of XXth Congress of the ISB.* (Cleveland, Ohio), pp. 120.

Dixon, S.J., Batt, M.E. and Collop, A.C., 1998, Artificial playing surfaces research: A review of medical, engineering and biomechanical aspects. *International Journal of Sports Medicine*, **20**, pp. 209-218.

FIFA, 2005, In *FIFA Quality Concept for Artificial Turf Guide* (Zurich: FIFA), pp. 1-43.

Junqua, A., Pavis, B., Lacouture, P., Niviere, J. and Rivat, A., 1983, About standards on sports floors. In *Biomechanical Aspects of Sports Shoes and Playing Surfaces*, edited by Nigg, B.M. (Calgary: University Press), pp. 77-82.

Nigg, B. M., 1983, External force measurements with sports shoes and playing surfaces. In *Biomechanical Aspects of Sports Shoes and Playing Surfaces*, edited by Nigg, B.M. (Calgary: University Press), pp. 11-23.

Nigg, B. M., 1990, The validity and relevance of tests used for the assessment of sports surfaces. *Medicine and Science in Sports and Exercise*, **22**, pp. 131-139.

UEFA, 2005, In *FIFA Quality Concept: Handbook of Test Methods and Requirements for Artificial Turf Football Surfaces* (Nyon: UEFA).

CHAPTER ELEVEN

Measurement error and global positioning systems for analysing work-rate

S. Post[1], A. P. Hollander[1] and T. Reilly[2]

[1]Department of Human Movement Sciences, Vrije Universiteit Amsterdam, Amsterdam, Netherlands
[2] Research Institute for Sport and Exercise Sciences, Liverpool John Moores University, Liverpool, UK

1. INTRODUCTION

Motion analysis has been the traditional tool for monitoring the work-rate of participants in soccer matches (Reilly and Thomas, 1976; Withers *et al.*, 1982; Bangsbo, 1994). Its principle is based on the total energy expended being dependent on the overall distance covered, irrespective of the velocity of movement. The original method was designed by Reilly and Thomas (1976) and validated against video recordings with the stride length for different activity categories being calibrated against reference observations for each individual. The method proved to be highly reliable and objective and yielded data that identified sources of fatigue, positional differences and relationships to physiological measures. The breakdown of movements into different categories for exercise intensity enabled later research groups to design laboratory-based protocols that simulated the intensity corresponding to match-play (Drust *et al.*, 1998). These work-rate profiles accommodated the contributions of both distance and velocity to the overall distance covered.

The dataset from motion analysis of players has permitted the close scrutiny of particular activities, for example a focus on the high-intensity component of the work-rate profile. As the methods evolved in technical sophistication, the data available became more comprehensive, concomitant observations being possible for all players in a game when multi-camera systems of monitoring movement were installed at the top professional clubs. The development of technologies for tracking players' movements, mostly video-based, have been reviewed by Carling and Williams (2008) and their applications to address training needs described elsewhere (Carling *et al.*, 2005). With sports science support teams now based in many of the top professional clubs in the major European leagues, the work-rate data are used primarily as feedback on playing performance. A limitation of the multi-camera system is that cost prohibits their installation in a training environment for any evaluation of training interventions.

Global positioning systems or GPSs are currently being promoted for tracking multiple participants simultaneously. The GPS unit worn by an individual records data on time, speed, distance, position, altitude and direction of motion. Heart rate may also be monitored but this requires the individual to wear a heart sensor strapped across the chest. Following exercise, the data are downloaded to a PC where further information is provided with respect to the breakdown of the intensity of locomotion (Edgecomb and Norton, 2006). The GPS sensor is worn by players in a lightweight back-pack positioned just below the back of the neck. A clear view of the sky provides optimal conditions for tracking individuals and so the instrumentation is unsuitable for indoor use but it still works well in bad weather.

Since the wearing of monitoring sensors is not permitted in competitive matches, the use of GPS systems for data acquisition is limited to training contexts and friendly matches. The utility of GPS has been apparent in outdoor activities where crude information rather than scrutiny of rapid changes in velocity is adequate. As the work-rate profile of players typically entails over one thousand changes in activity categories per game, soccer play imposes a much greater challenge for data collection than the linear locomotion sports such as cycling, running and outdoor sports (Larsson, 2003).

Validation attempts to authenticate GPS for research purposes have been cursory (Di Salvo *et al.*, 2006) or gone unreported in detail (Hewitt *et al.*, 2009), apart from the study of Australian Rules football players by Edgecomb and Norton (2006). Errors in recording of velocity and distance covered may vary between GPS units, particularly in short unorthodox types of activity such as occurs in playing soccer (Hebenbrock *et al.*, 2005).

The aim of this study was to validate the GPS system during soccer specific activities. There were two phases to the work engaged, first to compare the system with a reference method such as video recording and second, to determine the agreement between methods in monitoring activities during soccer training.

2. METHODS

2.1 Research design

The research programme consisted of two separate studies. In the first study, measurements were compared for subjects negotiating a running track using an intermittent exercise protocol. Subjects were recorded on video for calculation of distances covered and the data compared to the reference 400 m distance. The validity of video recording was checked by establishing the relative technical error of measurement. A consistent relative technical error would justify the video method if a correction factor could be applied to it. In the subsequent study, subjects were monitored during small-sided games using a GPS device, allowing results to be compared with those determined from conventional motion analysis of video recordings. The repeatability and objectivity of the reference method were examined in the course of this study. Both studies were approved by Liverpool John Moores University's Research Ethics Committee .

2.2 Study 1: validity of video analysis

Eight male university students (age 22.8 ± 3.5 years, body mass, 69.6 ± 3.9 kg, height 1.791 ± 0.067 m) participated in the study. Of the eight subjects, two were playing soccer on a regular basis. All eight subjects were fit and healthy and provided written informed consent.

Prior to validating the GPS, the relative technical error of the video method had to be calculated. To establish this relative technical error, subjects were video-filmed (camera) close up while running at a calibrated running track (400 m) doing different locomotor activities. Duplicate measurements were obtained for all subjects and the distance round the track was checked with a calibration wheel. The following activities were chosen, walk (w), jog (j), cruise (cr) and sprint (sp). These categories were chosen because they represent the activities covered during a soccer match (Reilly and Thomas, 1976). Mean stride lengths for all the activities and for every subject were determined by instructing the subjects to cover a distance (20 m) between two marked points. Using video analysis the mean stride lengths of all four activities were determined.

To cover the running track in a pragmatic way and be comparable to a soccer match, the running track was divided into 16 parts each of 25 m. Each 25-metres distance was pre-determined with a tape measure and was marked by cones. To incorporate the chosen activities (j, cr, w, sp) in a 400-m run, a running protocol was established. The protocol contained 4 similar blocks of (4*25 m) where the subjects ran the activities of j, cr, w, s without a pause, as illustrated in Figure 1. Total distance was determined by video analysis of the run and was calculated as the sum of distances covered during each type of activity. To establish reliability this total distance was compared with the actual length (400 m) of the running track. A consistent relative technical error would justify the video method if a correction factor could be applied to it.

2.3 Study 2: Comparison of methods

Eight youth soccer players of the Everton Academy (age 14.6 ± 0.5 years, body mass 64.3 ± 10.8 kg, height 1.698 ± 0.081 m) were studied. The players were observed individually by video filming them during a match or training. At the same time the player who was video-filmed was wearing a GPS receiver (The Sports Performance Indicator Elite, SPI elite, from GPSports Systems Pty. Ltd., http://www.gpsports.com). Sampling frequency was 1 Hz. The eight participants comprised one centre back, two left wingers, two right backs and three strikers.

Video-recordings were made during training activities and friendly matches on eight occasions. A video-camera (Sony TR 75E) followed a subject throughout parts of the training and the friendly matches. Data were burned on DVD and analysed on a computer (Targa vision XP). The discrete activities that were monitored were separated into four activity categories based on the intensity of action according to the classification of Drust *et al.* (1998). These activities were walking (forward, backward and sideways movements), jogging (forward, backward and sideways movements), cruising and sprinting. Static pauses were also recorded. For each discrete activity the number of strides was counted and converted to distance. Mean stride lengths were determined for every subject by covering a distance between two marked points (20 m) for each discrete activity. All these discrete activities were video-recorded so average stride length could be determined. During video analysis, the total distance covered during training activities and the friendly matches was calculated as the sum of the distances covered during each type of activity. Total distance obtained during the video analysis was compared with total distance monitored with a GPS device.

To assess inter-observer and intra-observer reliability (respectively, objectivity and repeatability) of the data monitored by means of the video-camera, some recordings were re-analysed. Repeatability was initially assessed three days after completion of the first analysis. To determine how well repeatability was maintained, the fourth and eighth matches were re-analysed as well. To obtain objectivity, another observer who had previous experience in analysing video recordings re-analysed one of the matches.

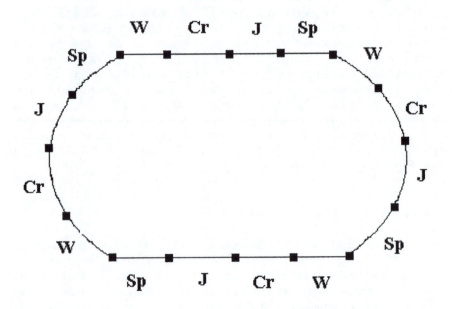

Figure 1. Outline of the running track with categories of activity denoted;
J = jog, W = walk, Cr = cruise, Sp = sprint.

2.4 Statistical analysis

The coefficient of variation (CV0) and 95% limits of agreement were used in the comparison of methods. For comparisons between the GPS and the video-method,

Bland and Altman's (1986) 95% limits of agreements were employed. The limits of agreements method is based on a graphical analysis of the mean of the GPS distances and the video distances versus the difference between GPS distances and the video distance. To obtain the 95% random error component, the Standard Deviation (SD) of the differences between the GPS method and the video method was first calculated and then multiplied by 1.96. The limits of agreements were than expressed as ± of this value. The mean of the limits of agreements is the mean difference between the GPS method and the video method and is expressed as the bias (Atkinson and Nevill, 1998). The absence of heteroscedasticity in the data was established, before the limits of agreements were calculated.

The repeatability and the objectivity of the total distance covered in each discrete activity were examined using coefficients of variation (CV) and limits of agreement (Bland and Altman, 1986). The exact agreements were examined using the kappa statistic of Altman (1991).

3. RESULTS and DISCUSSION

3.1 Study 1

Mean distance of the 16 video analysis was 381.83 (± 18.68) m and the actual distance of the running track was 399.20 m; this deviation from 400 m is considered small, accounted for by variations in the path of the trundle wheel from that used by track designers. The mean difference between methods was 17.37 m; to calculate relative technical error of the video measurement this mean difference was divided by the actual length of the running track and expressed as a percentage. The relative technical error for the video measurements was -4.35% (± 4.7). The video method had an average underestimation of 4.35%. The relative error of the video measurement was not consistent, so a correction factor could not be applied to the video data.

It seems that error is inherent in observation methods once the path of movement deviates from linearity. The subjects may use a path that economises on distance covered, especially when moving around the curves of the track. There is likely also to be error in the averaging approach used in calibrating stride length for each activity category. This error will be manifest when long runs are undertaken with stride lengths that exceed the average or when players accelerate, decelerate or change direction with unorthodox stride rates per unit distance. Error would be reduced if the shortening of strides in negotiating the curves were accounted for in calibrating stride lengths for the video analysis. Some reduction in error would be anticipated also if the calibration procedures were repeated, the subjects being available for one set of measurements only in this study.

3.2 Study 2

The kappa values for repeatability (k>0.84) and objectivity (k>0.74) corresponded to a strength of agreement considered to be 'very good' and 'good' respectively. The coefficient of variation was 5.7 (SD= 4.1)% for repeatability and 15.9 (SD=15.7)% for objectivity. These values for total distance covered and distance covered in the various categories of activity satisfy the criteria for measurement methods in biological systems.

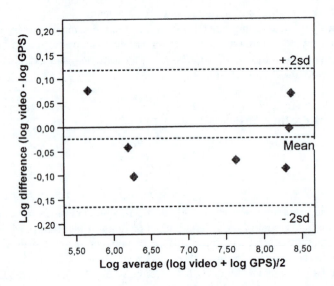

Figure 2. Bland-Altman plot for the average of the video and the GPS method transformed to logarithmic values against the difference between the GPS and the video method similarly transformed.

The mean distance covered in training and match scenarios was 2008 and 2029 m for the video analysis and GPS respectively, a mean difference of 1.03%. The correlation between methods was positive (r = 0.662; P< 0.05), indicating the presence of heteroscedasticity in the data (see Figure 2). The data were therefore transformed to logarithmic values before limits of agreement were calculated and transformed to anti-logs for obtaining ratio limits of agreement. Results for the comparison of the two methods indicated a mean error close to zero but the 95% LOA for GPS analysis ranged from 12% above to 15% below. These figures are very close to the values of -9.3 to +18.9% reported by Edgecomb and Norton (2006) who found a mean overestimation of 4.8 (7.2)% for the GPS method.

It seems that the GPS system is suitable for providing gross information about performance but loses accuracy when fine details are examined. The sampling frequency of 1 Hz is too low for acquiring accurate data on acceleration and fast changes in motion. Besides, Portas *et al.* (2007) have shown that the error in GPS increases with increasing velocity. Improvements in GPS technology, notably the incorporation of accelerometry and the use of higher sampling frequencies, should encourage the validity of GPS in a soccer context to be re-visited. The present comparison of methods might be replicated using alternative reference methods: these might include multi-camera systems or direct linear transformation which constructs two-dimensional positions on the pitch from a perspective-transformed image.

4. CONCLUSIONS

The conventional method of work-rate analysis by means of video recording was confirmed to be repeatable, objective and valid. Global positioning systems are replacing traditional recording in training contexts, due largely to their labour saving properties. Results from the present study justify the use of such tracking systems, although they are not yet sensitive to all of the movements occurring in a soccer game. Further validation studies are recommended as the quality of commercial systems is improved. Such studies should be conducted before GPS can be endorsed unequivocally for use by personnel working with soccer clubs in a sports science support role.

Acknowledgements

The authors are indebted for help with the organisation of this project provided by Phil Hewitt at Everton F.C.

References

Altman, D., 1991, *Practical Statistics for Medical Research* (London, United Kingdom: Chapman and Hall), pp. 402.

Atkinson, G. and Nevill, A.M., 1998, Statistical methods for assessing measurement error (reliability) in variables relevant to sports medicine. *Sports Medicine*, **2**, pp. 217-238.

Bangsbo, J. 1994, The physiology of soccer – with special reference to intermittent exercise. *Acta Physiologica Scandinavica*, **15**, pp. 1- 155.

Bland, M. and Altman, D.G., 1986, Statistical methods for assessing the difference between two methods of measurement. *Lancet* , **I**, pp. 307-310.

Carling, C. and Williams, A.M., 2008, Match analysis and elite soccer performance: combining science and practice. In *Science and Sports: Bridging the Gap*, edited by Reilly, T. (Maastricht: Shaker Publishing BV), pp. 32-47.

Carling, C., Williams, A.M. and Reilly, T., 2005, *Handbook of Soccer Match Analysis: A Systematic Approach to Improve Performance* (London: Routledge).

Di Salvo, V., Collins, A., McNeill, B. and Cardinale, M., 2006, Validation of

Prozone®: A new video-based performance analysis system. *International Journal of Performance Analysis in Sport*, **6**, pp. 108-119.

Drust, B., Reilly, T. and Rienzi, E., 1998, Analysis of work rate in soccer. *Sports Exercise and Injury*, **4**, pp. 151-155.

Edgecomb, S.J. and Norton, K.I., 2006, Comparison of global positioning and computer-based tracking systems for measuring player movement distance during Australian football. *Journal of Science and Medicine in Sport*, **9**, pp. 25-32.

Hebenbrock, M., Due, M., Holzhausen, H., Sass, A., Stadler, P. and Ellendorff, F., 2005, A new tool to monitor training and performance of sport horses using global positioning system (GPS) with integrated GSM capabilities. *Deutsche Tierarztliche Wochenschrift*, **112**, pp. 262-265.

Hewitt, A., Withers, R. and Lyons, K., 2009, Match analysis of Australian international female soccer players using an athlete tracking device. In *Science and Football VI*, edited by Reilly, T. and Korkusuz, F. (London: Routledge), pp. 224-228.

Larsson, P., 2003, Global positioning system and sport-specific testing. *Sports Medicine*, **33**, pp. 1093-101.

Portas, M., Rush, C., Barnes, C. and Batterham, A., 2007, Method comparison of linear distance and velocity measurements with global positioning satellite (GPS) and the timing gate techniques. *Journal of Sports Science and Medicine*, **10**, pp. 001-009.

Reilly, T. and Thomas, V., 1976, A motion analysis of work rate in different positional roles in professional soccer match-play. *Journal of Human Movement Studies*, **2**, pp. 87-97.

Withers, R.T., Maricic, Z., Wasilewski, S. and Kelly, L., 1982, Match analysis of Australian professional soccer players. *Journal of Human Movement Studies*, **8**, pp. 159-76.

Part III
Physiology and Medicine

Effect of soccer-specific fatigue on eccentric hamstring strength: Implications for hamstring injury risk

K. Small[1], L. McNaughton[1], M. Greig[2] and R. Lovell[1]

[1]Department of Sport, Health and Exercise Science,
University of Hull, Hull, England
[2]The Football Association, Lilleshall NSC, Shropshire, England

1. INTRODUCTION

Recent epidemiological studies have documented an increased number of hamstring strain injuries, which are now frequently reported as the most common injury to soccer players (Hawkins and Fuller, 1999; Hawkins *et al.*, 2001; Árnason *et al.*, 1996, 2004; Woods *et al.*, 2004). They are well recognised by medical personnel, coaches and athletes as a major cause for concern as well as the substantial associated financial costs and consequences for inhibiting performance (Woods *et al.*, 2004). In order to minimise the incidence of hamstring strains and associated costs, more effective intervention programmes for injury prevention are recommended (Rahnama *et al.*, 2002), prior to which a greater understanding is required to help determine the aetiology and mechanisms of injury.

Fatigue has been associated as an aetiological risk factor for hamstring injuries, with soccer injury audits revealing that almost half (47%) of all hamstring injuries during matches occur during the last third of the first and second halves (Woods *et al.*, 2004). This observation supports the notion that fatigue is a predisposing factor to hamstring strain injury. Muscle strength deficiency has furthermore been proposed as a factor affected by fatigue to increase susceptibility to hamstring injury (Greig, 2008). A significant decrease in eccentric hamstring strength (Gleeson *et al.*, 1998), combined with reduced functional eccentric hamstrings: concentric quadriceps strength ratio has been observed during simulated soccer match-play (Rahnama *et al.*, 2003; Greig, 2008).

Previous research into the effect of fatigue associated with soccer match-play on muscle strength has employed unidirectional protocols using a motorised treadmill (Rahnama *et al.*, 2003; Greig, 2008), thus not reflecting the multidirectional nature of soccer, or employed a protocol not based on actual match-play to support the duration, speed or intensity of activity (Gleeson *et al.*, 1998). Hence, the aims of this study were to investigate the effect of a 90-min multidirectional free-running soccer-specific fatigue protocol based on contemporary match-play data (Prozone®) on hamstring and quadriceps strength and muscular imbalances and establish the implications of this type of fatigue for injury risk to the hamstrings.

2. METHODS

2.1 Participants

Sixteen, uninjured, male, semi-professional soccer players (Mean ± SD; Age: 21.3 ± 2.9 years; height 1.85 ± 0.08 m; body mass 81.6 ± 6.7 kg) took part in this investigation. Written, informed consent was obtained prior to data collection from the participants, and ethical approval for the study obtained in accordance with the departmental and university ethical procedures.

2.2 Experimental design

Participants completed a 90-min Soccer-specific Aerobic Field Test (SAFT[90]), divided into two 45-min periods. Prior to exercise (t0), at half-time (t45) and post-exercise (t105), participants performed three maximal isokinetic movements of the dominant limb for concentric knee extensors (conQ), concentric knee flexors (conH) and eccentric knee flexors (eccH).

2.3 Experimental protocol

The SAFT[90] was developed to replicate the physiological and mechanical fatigue reflective of soccer match-play, and was based on contemporary time-motion analysis data obtained from 2007 English Championship league matches (Prozone[®]). The test was designed to include multidirectional and utility movements, with frequent acceleration and deceleration as is inherent to soccer. The design of the course was based around a shuttle run over a 20-m distance, with the incorporation of four positioned poles for the participants to navigate using utility movements.

   A 15-min activity profile was developed using the match-play data (Figure 1), which was completed six times during the full 90-min simulated soccer match activity, with a passive rest interval of 15 min for half-time. The activity profile was characterised by standing (0.0 km·h⁻¹), walking (4.0 km·h⁻¹), jogging (10.3 km·h⁻¹), striding (15.0 km·h⁻¹) and sprinting (≥ 20.4 km·h⁻¹) in a randomised and intermittent fashion dictated by verbal signals via an audio CD. Players covered a total of 10.8 km with 1269 changes in speed (every 4.3 s), and 1350 changes in direction over the 90 min.

2.4 Muscle strength profiling

Isokinetic peak torque for the knee flexors and extensors (of the participants' dominant leg; their 'kicking' leg) was measured using an isokinetic dynamometer

(Biodex System 3). Prior to all testing, participants performed a standardised warm-up procedure involving 5 min on a cycle ergometer at between 50-60 rev·min⁻¹, 5 min static and dynamic stretches for the major lower-limb muscle groups and 5 min light jogging and familiarisation with the SAFT⁹⁰ exercise protocol.

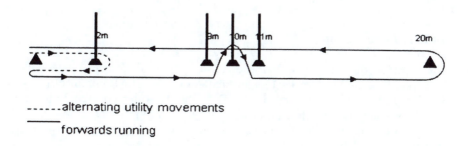

------alternating utility movements

——— forwards running

Figure 1. Diagrammatic representation of a 15-minute activity profile of the SAFT⁹⁰.

During testing, participants were seated on the dynamometer in an adjustable chair with straps secured to stabilise their body. Participants initially performed two practice sub-maximal knee flexion and extension movements followed by three maximal efforts for each muscle action. The order of testing was standardised (concentric knee extension followed by concentric knee flexion and eccentric knee extension) for subsequent testing throughout the SAFT⁹⁰ protocol, at half-time and post-exercise, with three trials completed in total. All actions were performed at an angular velocity of 2.09 rad·s⁻¹ (120°·s⁻¹) and through a range of 0° to 90° knee flexion and extension (with 0° being full knee extension).

Gravity-corrected peak torque values were extracted from the strength indices as comparable measures between participants of muscular performance, as well as angles of peak torque. Mean peak torque values were calculated from the three attempts.

Using the recorded data, eight primary variables were selected for further analysis: concentric quadriceps peak torque (PT), concentric hamstring PT, eccentric hamstring PT, the traditional concentric hamstrings: concentric quadriceps strength ratio, the functional eccentric hamstrings: concentric quadriceps strength ratio, concentric quadriceps angle of peak torque (APT), concentric hamstrings APT and eccentric hamstrings APT.

2.5 Statistical analysis

Descriptive statistics were initially used to calculate means and standard deviations. An analysis of variance for repeated measures, with the least significant difference (LSD) multiple comparison procedure and post-hoc test was then used to compare results from muscle strength variables measured at each of the three time points: prior-to exercise (t0), at half-time (t45) and post-exercise (t105). Statistical analysis of the data collected was processed using SPSS statistical software with significance levels set at $P \leq 0.05$.

3. RESULTS

3.1 Peak torque results

The repeated measures ANOVA showed a significant decrease in eccentric hamstring peak torque (eccH PT) with time during the SAFT[90] protocol ($F_{2,30}$ = 3.6, $P < 0.001$; Partial Eta squared = 0.672) (Table 1), with an overall decrease of 16.8% over the 90 min. No significant changes with fatigue during the SAFT[90] protocol were observed for the concentric quadriceps and concentric hamstrings peak torque ($P > 0.05$).

Table 1. Mean (±SD) concentric quadriceps peak torque (conQ PT), concentric hamstring peak torque (conH PT) and eccentric hamstring peak torque (eccH PT) over time. *Significant difference between t0 and t105 ($P < 0.05$). †Significant difference between t0 and t45 ($P < 0.05$). ‡Significant difference between t45 and t105 ($P < 0.05$).

Time (min)	conQ PT (N·m)		conH PT (N·m)		eccH PT (N·m)	
	Mean	±SD	Mean	±SD	Mean	±SD
0	235.79	19.74	140.41	36.87	271.95 * †	43.15
45	225.69	21.93	134.28	27.08	240.40 ‡ †	43.24
105	229.12	18.46	132.05	20.33	226.31 * ‡	45.66

3.2 Strength ratio results

The ANOVA revealed a significant reduction in the functional eccH:conQ strength ratio during the SAFT[90] ($F_{2,30}$ = 9.9, $P = 0.001$; Partial Eta squared = 0.398) (Table 2), declining by 14.95% over the 90-min protocol. No significant change was observed during the SAFT[90] for the traditional concentric hamstrings:concentric quadriceps strength ratio ($P > 0.05$).

3.3 Angles of peak torque

Repeated measures ANOVA revealed significant changes in the conQ angle of peak torque during the SAFT[90] ($F_{2,30}$ = 4.1, $P < 0.03$; Partial Eta squared = 0.214), with a trend towards angle of peak torque at a longer muscle length during the

SAFT[90]. The conH angle of peak torque also demonstrated a shift towards a longer muscle during the SAFT[90], with ANOVA revealing significant changes over time ($F_{2,30} = 6.3$, $P = 0.005$; Partial Eta squared = 0.297). These results indicate a shift in the APT for conQ and conH towards longer muscle lengths with fatigue during the later stages of each half of the SAFT[90] (Table 3).

A repeated measures ANOVA revealed significant changes in eccH angle of peak torque with time ($F_{1.37,20.58} = 4.4$, $P < 0.04$; Partial Eta squared = 0.227). Contrastingly, these results indicated a shift in the angle of peak torque for eccH towards shorter muscle lengths during the SAFT[90] (Table 3).

Table 2. Mean (±SD) traditional concentric hamstrings:concentric quadriceps strength ratio (trad H:Q) and functional eccentric hamstrings:concentric quadriceps strength ratio (func H:Q) over time.
* Significant difference between t0 and t105 ($P < 0.05$).
† Significant difference between t0 and t45 ($P < 0.05$).

Time	Trad H:Q (%)		Func H:Q (%)	
(min)	Mean	±SD	Mean	±SD
0	59.47	14.53	115.90 * †	20.70
45	60.29	11.19	106.62 †	17.18
105	57.81	9.60	98.92 *	19.68

Table 3. Mean (±SD) angle of peak torque (APT) for concentric quadriceps (conQ), concentric hamstrings (conH) and eccentric hamstrings (eccH) over time. (0° = full knee extension).
*Significant difference between t0 and t105 ($P < 0.05$).
† Significant difference between t0 and t45 ($P < 0.05$).

Time	conQ APT (°)		conH APT (°)		eccH APT (°)	
(min)	Mean	±SD	Mean	±SD	Mean	±SD
0	71.8 * †	6.1	65.1 * †	12.1	28.2 * †	11.7
45	75.9 †	6.2	54.3 †	13.5	36.7 †	14.5
105	75.5 *	4.9	51.9 *	14.0	38.2 *	18.2

4. DISCUSSION

The SAFT[90] exercise protocol designed to replicate the physiological and mechanical demands of soccer match-play activity was shown to induce a diminished capacity of the knee flexor muscles to generate eccentric force. Consequently, the SAFT[90] also impaired the functional eccH:conQ strength ratio and muscular balance.

Findings were principally related to a reduction in eccentric hamstring peak torque which decreased by 16.8% by the end of the 90-min multidirectional soccer simulation compared with pre-exercise values. This result supports previous research findings, with Rahnama *et al.* (2003) revealing an identical decrement in eccentric hamstring muscle peak torque; 16.8% (using 2.09 rad·s⁻¹ angular velocity), during a 90-min soccer fatiguing protocol performed on a motorised treadmill using allied muscle strength profiling. The functional eccH:conQ strength ratio during the two exercise protocols also diminished to a similar extent, with a reduction of 15.0% recorded during the current study after completing the SAFT[90], and 13.0% observed following the treadmill based protocol employed by Rahnama *et al.* (2003).

It is clear that the two exercise protocols vary substantially in their activity profile and nature. The SAFT[90] involves forwards, backwards, sideways and cutting actions throughout the course, whilst the activity profile incorporates 1269 transitions in speed (on average once every 4.3 s) and 1350 changes in direction within the full 90-min protocol. In contrast, the protocol employed by Rahnama *et al.* (2003) was that developed by Drust *et al.* (2000) and performed on a programmable treadmill. Due to technical limitations of the equipment, no utility movements were performed, and the activity profile contained only 92 discrete bouts of activity during the full 90 min, with longer time spent performing individual bouts (walking 35.3 s, jogging 50.3 s, cruising 51.4 s and sprinting 10.5 s). Increased time spent performing the high-intensity cruising and sprinting in particular, totalling over 50% of total distance covered compared with ~14% observed during matches (Thatcher and Batterham, 2004), could have created the additional muscular requirement to match the extra load imposed by the multidirectional and more frequently intermittent SAFT[90] protocol. Consequently, the results from the current study may be more representative of the response associated with activity during actual soccer match-play. Alternatively, it could be concluded that the type of exercise is not essential for representing the exercise intensity of soccer and that fatigue occurs independently over 90 min of activity. Therefore, the results reported by Rahnama *et al.* (2003) can be generalised.

Reduced eccentric hamstring strength is commonly associated with hamstring strain injury risk, as it is thought that during powerful eccentric contractions, muscles are most susceptible to injury (Garrett, 1996; Verrall *et al.*, 2001; Brockett *et al.*, 2004). Reduced eccentric hamstring peak torque was also observed by Gleeson *et al.* (1998) following a 90-min free-running soccer fatiguing protocol accompanied by a reduced knee joint stability and an increased risk of hamstring injury. This observation could furthermore relate to the reduction in the functional eccH:conQ strength ratio observed during the SAFT[90]. Therefore, the decline in peak eccentric hamstring torque and functional eccH:conQ strength ratio with fatigue during the SAFT[90] may help explain the increased predisposition to hamstring strain injury during the later stages of matches observed by Woods *et al.* (2004). Hamstring injury prevention strategies should therefore aim to develop strengthening programmes that can maintain eccentric hamstring strength and preserve the functional eccH:conQ strength ratio during soccer match-play. This could involve altering the timing of when eccentric hamstring strengthening exercises are employed during training from the traditional approach of when non-fatigued, to a new strategy of strength training in fatigued conditions. Regarding the law of specificity (Kraemer *et al.*, 2002), training in a fatigued state could improve performance in a fatigued state thus better maintaining muscle strength. This approach could have beneficial effects and important implications for prevention of hamstring injuries.

There was no significant reduction observed in the concentric quadriceps or hamstrings peak torque during the SAFT[90], which although contradicting previous results of Rahnama *et al.* (2003) may again be explained by differences in the exercise protocols employed. However, the respective changes in the angle of peak torque for the concentric quadriceps and hamstrings muscle actions did reveal a shift in the optimum length for peak muscle tension in the direction of longer muscle lengths with fatigue during the SAFT[90]. This finding might imply that the protocol employed by Rahnama *et al.* (2003) caused a quantitative change in concentric quadriceps and hamstrings peak torque, whereas the SAFT[90] produced a more qualitative shift in angle of peak torque. Various theories have been proposed to help explain the shift in angle of peak torque for the concentric quadriceps and hamstrings muscle actions observed in the present study, with perhaps the most widely accepted being the "popping sarcomere hypothesis" as proposed by Morgan (1990). This proposes that with fatigue, micro-tears occur within a muscle with effective compliance of the muscle fibre increasing, leading to a shift of the whole muscle's length-tension relationship towards longer lengths. A significantly greater loss of relative force at shorter muscle length has been associated with muscle damage from eccentric contractions (Byrne *et al.*, 2001), and therefore risk of injury (Brockett *et al.*, 2001).

The eccentric hamstrings angle of peak torque was also significantly altered with fatigue during the exercise protocol, although results contrastingly revealed a shift towards a shorter muscle length with fatigue. Whilst research into this area is in part equivocal, the results support findings by Proske *et al.* (2004), who examined values for torque-angle curves for human hamstring muscles obtained before and immediately after a series of eccentric hamstring knee extensions on an isokinetic dynamometer. Results indicated a shift in the length-tension curve towards shorter muscle lengths following the fatigue protocol in agreement with findings from the current investigation.

This finding may have important implications for injury risk, as it is believed that athletes who produce peak torque at shorter muscle lengths are at a greater risk of injury (Brockett *et al.*, 2001; Brockett *et al.*, 2004; Proske *et al.*, 2004). Furthermore, it has been proposed that peak torque generation at a shorter optimum muscle length would mean more of the muscle's operating range would be on the descending limb of the length-tension curve (Brughelli and Cronin, 2007). Therefore, as muscles are more likely to become injured when operating in a more lengthened position (Garrett, 1990), this could explain the increased predisposition to hamstring strain injuries during the later stages of matches when the angle of peak torque has been observed to be at a shorter muscle length, and especially with the concurrent deterioration in eccentric hamstring peak torque and reduced functional eccH:conQ strength ratio also reported.

5. CONCLUSIONS

The SAFT[90] exercise protocol produced a time dependent decrease in eccentric strength of the knee flexors, and consequently in the functional eccH:conQ strength ratio. Additionally, the multidirectional, soccer-specific fatigue protocol produced a shift in the angle of peak torque towards longer muscle lengths during concentric knee flexor and extensor actions with time, whereas a shift towards shorter muscle length was observed with fatigue during the eccentric knee extensor action.

These findings have implications for the increased predisposition to hamstring strain injury during the later stages of soccer matches. Future researchers, club trainers and medical staff should attempt to take into account this fatigue effect associated with soccer match-play when designing programmes for preventing hamstring injuries. Strategies that can reduce the negative effects of fatigue during match-play, such as strength training in a fatigued state, may then help lower the risk of hamstring injury when fatigued during soccer matches.

References

Árnason, A., Gudmumdsson, A., Dahl, H.A. and Johannsson, E., 1996, Soccer injuries in Iceland. *Scandinavian Journal of Medicine and Science in Sports*, **6**, pp. 40-45.

Árnason, A., Sigurdsson, S.B., Gudmundsson, A., Holme, I., Engebretsen, L. and Bahr, R., 2004, Risk factors for injuries in football. *American Journal of Sports Medicine*, **32**, 5S-16S.

Brockett, C.L., Morgan, D.L. and Proske, U., 2001, Human hamstring muscles adapt to eccentric exercise by changing optimum length. *Journal of Medicine and Science in Sports and Exercise*, **33**, pp. 783-790.

Brockett, C.L., Morgan, D.L. and Proske, U., 2004, Predicting hamstring strain injury in elite athletes. *Journal of Medicine and Science in Sports and Exercise*, **36**, pp. 379-387.

Brughelli, M. and Cronin, J., 2007, Altering the length-tension relationship with eccentric exercise. *Journal of Sports Medicine and Physical Fitness*, **37**, pp. 807-826.

Byrne, C., Eston, R.G. and Edwards, R.H.T., 2001, Characteristics of isometric and dynamic isometric strength loss following eccentric exercise-induced muscle damage. *Scandinavian Journal of Medicine and Science in Sports*, **11**, pp. 134-140.

Drust, B., Reilly, T. and Cable, N.T., 2000, Physiological responses to laboratory-based soccer-specific intermittent and continuous exercise. *Journal of Sports Sciences*, **18**, pp. 885-892.

Garrett, W.E. Jr., 1990, Muscle strain injuries: Clinical and basic aspects. *Medicine and Science in Sports and Exercise*, **22**, pp. 436-443.

Garrett, W.E. Jr., 1996, Muscle strain injuries. *American Journal of Sports Medicine*, **28**, S2-S8.

Gleeson, N.G., Reilly, T., Mercer, T.H., Rakowski, S. and Rees, D., 1998,

Influence of acute endurance activity on leg neuromuscular and musculoskeletal performance. *Medicine and Science in Sports and Exercise,* **30**, pp. 596-608.

Greig, M., 2008, The influence of soccer-specific fatigue on peak isokinetic torque production of the knee flexors and extensors. *American Journal of Sports Medicine,* in press.

Hawkins, R. and Fuller, C.W., 1999, A prospective epidemiology study of injuries in four English professional football clubs. *British Journal of Sports Medicine,* **33**, pp.196-203.

Hawkins, R.D., Hulse, M.A., Wilkinson, C., Hodson, A. and Gibson, M., 2001, The association football medical research programme: an audit of injuries in professional football. *British Journal of Sports Medicine,* **35**, pp. 43-47.

Kraemer, W.J., Adams, K., Cafarelli, E., Dudley, G.A., Dooly, C., Feigenbaum, M.S., Fleck, S.J., Franklin, A.C., Hoffman, J.R., Newton, R.U., Potteiger, J., Stone, M.H., Ratamess, N.A. and Triplett-McBride, T., 2002, Joint position statement: progression models in resistance training for healthy adults. *Medicine and Science in Sports and Exercise,* **34**, pp. 364-380.

Morgan, D.L., 1990, New insights into the behaviour of muscle during active lengthening. *Biophysical Journal,* **57**, pp. 209-221.

Proske, U., Morgan, D.L., Brockett, C.L. and Percival, P., 2004, Identifying athletes at risk of hamstring strains and how to protect them. *Medicine and Science in Sports and Exercise,* **36**, pp. 379-387.

Rahnama, N., Reilly, T. and Lees, A., 2002, Injury risk associated with playing actions during competition soccer. *British Journal of Sports Medicine,* **36**, pp. 354-359.

Rahnama, N., Reilly, T., Lees, A. and Graham-Smith, P., 2003, Muscle fatigue induced by exercise simulating the work rate of competitive soccer. *Journal of Sports Sciences,* **21**, pp. 933-942.

Thatcher, R. and Batterham, A.M., 2004, Development and validation of a sport-specific exercise protocol for elite and youth soccer players. *Journal of Sports Medicine and Physical Fitness,* **44**, pp. 15-22.

Verrall, G.M., Slavotinek, J.P., Barnes, P.G., Fon, G.T. and Spriggins, A.J., 2001, Clinical risk factors for hamstring muscle strain injury: a prospective study with correlation of injury by magnetic resonance imaging. *British Journal of Sports Medicine,* **35**, pp. 435-439.

Woods, C., Hawkins, R.D., Maltby, S., Hulse, M., Thomas, A. and Hodson, A., 2004, The Football Association Medical Research Programme: an audit of injuries in professional football - analysis of hamstring injuries. *British Journal of Sports Medicine,* **38**, pp. 36-41.

CHAPTER THIRTEEN

Influence of one vs two matches a week on physiological and psychological profile of sub-elite footballers

I. Rollo[1], T. Curtis[1], L. Hunter[1] and F. Marcello Iaia[2]

[1]School of Sport and Exercise Sciences, Loughborough University, Loughborough, UK
[2]Faculty of Exercise Sciences, State University of Milan, Milan, Italy

1. INTRODUCTION

English soccer is home to one of the most demanding leagues in the world, the Premiership. In a typical season a player may be required to play close to forty games over a nine-month period, excluding any number of international and cup commitments. Due to the congestion of fixtures and pressure to achieve results, players are often required to play two competitive matches in a single week.

It is well established that fatigue occurs during a football game (Mohr *et al.*, 2003) and that various physiological resources take a few days to be completely restored. Specifically, neuromuscular fatigue and muscle soreness may compromise the players' ability to perform jumps and repeated sprints (Andersson *et al.*, 2008). In addition, poor dietary choices may prevent the restoration of pre-match glycogen concentrations (Jacobs *et al.*, 1982), required if the player is to perform prolonged high-intensity intermittent exercise.

At present there is no information available regarding the long-term influence of playing multiple games in a week on the recovery process and subsequent performance in soccer. Therefore, the aim of this study was to monitor the physiological and psychological profiles of 30 sub-elite soccer players over a six-week period. The squad of players was exposed to the same relative training load i.e. the players attended the same training sessions over the six-week period. The only difference was that half the players played one competitive game per week (ONE-GAME group) and the other half played two competitive games per week (TWO-GAME group).

2. METHODS

2.1 Participants and procedures

Thirty, healthy, experienced male soccer players (30±10 years, body mass 76.7.0 ±

1.4 kg, height 1.80 ± 0.02 m) (mean $\pm$ SEM) took part in the study.

A parallel two-group (ONE vs. TWO-GAME), longitudinal (0, 3, 6 weeks) design was used. On three separate occasions, 0, 3 and 6 weeks, 48 h after the last match, players completed a series of physiological tests in an indoor sports hall (at $18°$C, 50% relative humidity). All players were fully habituated with the physiological tests, completed in the same order (1, counter-movement jump, 2, 10-m and 20-m sprints, 3, Yo-Yo IRT1) and by the same experienced test leader. The Yo-Yo test IRT1 has been shown to have a coefficient of variation (CV) of approximately 8% (Bangsbo *et al.*, 2008). The CV of jump and sprint performance is approximately 2.4% (Markovic *et al.*, 2004). Each player was also provided with a questionnaire (Rest-Q 52) that was completed prior to the first training session of each week.

2.2 Statistics

Student's unpaired t-tests were used to compare subjects' characteristics between the two groups before the experimental period. Exercise performances were analysed using a two-factor repeated measures analysis of variance (RM ANOVA). If a significant interaction was detected, a Newman-Keuls post-hoc test was subsequently applied to locate the differences.

3. RESULTS

The TWO-GAME group showed decreased ($P<0.05$) CMJ (37.9 ± 5.5 vs. 35.0 ± 5.0 cm) (mean $\pm$ SEM), 10 m (1.80 ± 0.1 vs. 1.85 ± 0.1 s) and 20 m (3.08 ± 0.1 vs. 3.18 ± 0.1 s) sprint performances from week 3 to week 6. In the ONE-GAME group, CMJ and sprint performances were improved ($P<0.05$) by 10.7 and 3.0% respectively over the 6-week period. In the TWO-GAME group, at week 6 the Yo-Yo IR1 performance was ~11% lower ($P<0.05$) than at weeks 0 and 3 whereas no changes were observed in the ONE-GAME group (Figure 1). At weeks 3 and 6, in the REST-Q 52, scores of emotional and social stress, lack of energy and general well-being were worse ($P<0.05$) in the TWO-GAME as compared to the ONE-GAME group. No differences between ONE and TWO-GAME groups were observed for the other psychological variables.

4. DISCUSSION

The main finding of the present study was that the players` ability to jump, sprint and perform repeated high-intensity exercise was impaired when playing two competitive matches a week over 6 weeks. This was not the case for players in the ONE-GAME group, who actually improved jump and sprint scores by week 6.

An interesting observation is that none of the physical capabilities was impaired when playing two games per week over a 3-week period. This suggests that for 3 weeks, physical performance could be maintained despite playing an elevated number of matches. Instead there appears to be an accumulative effect of playing twice a week for 6 weeks on physiological performance. The scientific explanation of such an effect is likely to be ascribed to exacerbation in a number of factors including muscle fibre damage (Andersson *et al.*, 2008), non-optimal replenishment of muscle glycogen stores (Andersson *et al.*, 2008) and alterations in properties of the neuromuscular system (Andersson *et al.*, 2008).

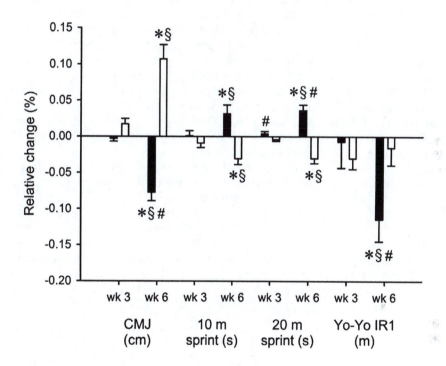

Figure 1. Effect of playing ONE (□) vs TWO (■) competitive games per week over 6 weeks on exercise performance. Values are means ± SE. * Significantly ($P<0.05$) different from wk 0. § Significantly ($P<0.05$) different from wk 3. # Significantly ($P<0.05$) different from playing ONE game per week.

Practically, the findings of the present study should stimulate clubs to monitor the physiological load which is placed upon players over a season and over periods of fixture congestion. Clubs should also be encouraged to evaluate current strategies in physiological and psychological recovery, to ensure maintained or enhanced player performance (Nicholas *et al.*, 1999; Williams and Serratosa, 2006).

In conclusion, players' ability to sprint, jump and perform repeated intense exercise was impaired when playing competitive matches twice a week over a 6-week but not a 3-week period. In addition, several psychological measures were negatively affected by playing two times per week as compared to once per week.

References

Andersson, H., Raastad, T., Nilsson, J., Paulsen, G., Garthe, I. and Kadi, F., 2008, Neuromuscular fatigue and recovery in elite female soccer: effects of active recovery, *Medicine and Science in Sports and Exercise*, **40**, pp. 372-380.

Bangsbo, J., Iaia, F.M. and Krustrup, P., 2008, The Yo-Yo intermittent recovery test - a useful tool for evaluation of physical performance in intermittent sports. *Sports Medicine*, **38**, pp. 37-51.

Jacobs, I., Westlin, N., Karlsson, J., Rasmusson, M. and Houghton, B., 1982, Muscle glycogen and diet in elite soccer players. *European Journal of Applied Physiology and Occupational Physiology*, **48**, pp. 297-302.

Markovic, G., Drazan, D., Jukic, I. and Cardinale, M., 2004, Reliability and factorial validity of squat and counter-movement jump tests. *Journal of Strength and Conditioning Research*, **18**, pp. 551–555.

Mohr, M., Krustrup, P. and Bangsbo, J., 2003, Match performance of high-standard soccer players with special reference to development of fatigue. *Journal of Sports Sciences*, **21**, pp. 519-528.

Nicholas, C. W., Tsintzas, K., Boobis, L. and Williams, C., 1999, Carbohydrate-electrolyte ingestion during intermittent high-intensity running. *Medicine and Science in Sports and Exercise*, **31**, pp. 1280-1286.

Williams, C. and Serratosa, L., 2006, Nutrition on match day. *Journal of Sports Sciences*, **24**, pp. 687-697.

CHAPTER FOURTEEN

The effect of an intense period of fixtures on salivary cortisol and IgA concentrations in professional soccer players

N.D. Clarke, A. Blanchfield, B. Drust, D.P.M. MacLaren and T. Reilly

Research Institute for Sport and Exercise Sciences,
Liverpool John Moores University, Liverpool, UK

1. INTRODUCTION

The competitive soccer calendar incorporates matches that vary in frequency from once a week to periods when 3 games can be played in 8 days. Repeated competitive strain has been expressed as a concern in the public domain, with some professional coaches seeking support for a mid-season break. From both physiological and administrative viewpoints the logic appears flawed, since an intermission would displace matches into another part of the season and cause further congestion of fixtures with an accompanying strain. Furthermore, a retrospective analysis of competitive engagements of players prior to international tournaments (Ekstrand et al., 2004) has highlighted that a congestion of fixtures rather than absence of a mid-season break is the more likely cause of players underperforming.

Several studies have shown that various aspects of immune function and humoral responses are temporarily suppressed after high-intensity exercise (Pedersen, 1991; Shephard et al., 1994). The period post-exercise where aspects of the immune system are suppressed may increase susceptibility to upper respiratory tract infections (URTI) and has been referred to as the 'open window' (Nieman, 1994). Cortisol, which is described as an immunosuppressive and anti-inflammatory agent (Weicker and Werle, 1991), has been shown to increase following intense exercise (Chicharro et al., 1998). In addition, a small progressive reduction in salivary IgA has been observed following repeated bouts of intermittent exercise (Sari-Sarraf et al., 2008). It has also been reported that salivary immunoglobulin A (sIgA) is decreased in 'over-trained' subjects (Pedersen et al., 2001).

Cortisol plays a major role in metabolism and immune function and is often used as an indicator of training stress. After intense exercise, it has been reported that there is an acute increase in cortisol levels (Filaire et al., 2001b). In addition a soccer training programme has been shown to increase salivary cortisol levels (Filaire et al., 2001a). However, the effect of training on resting levels of salivary cortisol is still unclear, with some studies showing no effect (Filaire et al., 1996) and some others showing either an increase (Seidman et al., 1990) or a decrease

(Tabata *et al.*, 1990).

Salivary IgA provides a major defence against potential pathogens by preventing colonization and replication on the mucosal surfaces of the upper respiratory tract. A reduction in the concentration of sIgA may allow for increased pathogenesis via the mucosal surface (Lamm, 1997). Subsequently it has been suggested that a reduction in sIgA is a possible reason behind the increased susceptibility of athletes to upper respiratory tract infections (URTI) (Nieman and Nehlsen-Cannarella, 1991). A seven-month training programme in elite swimmers has been found to decrease sIgA levels which were inversely correlated with the number of URTI (Gleeson *et al.*, 1999). However, sIgA levels have also been shown to remain constant (Mackinnon and Hooper, 1994) or increase (Gleeson *et al.*, 2000) in swimmers during a training season. The acute effects of exercise on sIgA are also unclear with reports of sIgA levels being decreased (Mackinnon *et al.*, 1993), increased (Sari-Sarraf *et al.*, 2007), or remaining unchanged (Bishop *et al.*, 2000) immediately following prolonged physical activity. The inconsistency in these findings may be explained by differences in fitness level of subjects, exercise protocols, saliva collection or storage methods, and the method used to express sIgA (i.e. whether sIgA was reported as absolute concentration, secretion rate, ratio to total saliva protein, or ratio to saliva osmolality) (Walsh *et al.*, 2002).

Changes in sIgA concentrations have been reported in elite female soccer players during a competition period (Akimoto *et al.*, 2003) and coincide with or precede the appearance of URTI in collegiate soccer players (Nakamura *et al.*, 2006). The higher incidence of infection reported in elite performers may be as a consequence of frequent high-intensity exercise, with little time for recovery in between. This schedule is similar to what occurs during busy periods in the competitive soccer calendar. The aim of the present study was to monitor sIgA and cortisol concentrations in professional players during an intense period of fixtures (3 matches played in 8 days).

2. METHODS

During an intense period of fixtures (3 matches in 8 days), saliva samples were collected from the outfield professional players at an English club competing in the 2007-2008 Coca-Cola Championship within 10 min of completing each match (the team gained promotion to the English Premier League at the end of the season). Players who had not participated in all of the matches or had been substituted were excluded; therefore the sample size was restricted to 6 players (mean age: 24 ± 1 years; height: 1.82 ± 0.0 m; body mass: 74.3 ± 3 kg). The 'passive expectoration method' was used according to the directions given by Navazesh and Christensen (1982). Initially, participants were required to rinse out their mouths with distilled water to prevent potential sample contamination that might affect sIgA and/or cortisol levels. Whole unstimulated saliva was collected by expectoration for 5 min into sterile pre-weighed plastic containers (Sarstedt, UK) with the subject's eyes open, head tilted slightly forward and making minimal orofacial movement. Saliva was weighed to the nearest mg and volume calculated assuming a saliva density of

1.00 g·ml⁻¹. The saliva was frozen and subsequently analyzed for cortisol (Salimertics, USA) and sIgA (Salimertics, USA) using a sandwich ELISA method.

Statistical analysis was conducted using one-way ANOVAs. All results are reported as the mean ± the standard error of the mean (SEM) and a level of $P<0.05$ was considered statistically significant.

3. RESULTS

Salivary cortisol concentration (Figure 1) was not different between match 1 (0.42 ± 0.1 hg·ml⁻¹) and match 2 (0.42 ± 0.2 hg·ml⁻¹). Following the third match the concentration of salivary cortisol had increased to 0.63 ± 0.2 hg·ml⁻¹ although this difference was revealed not to be significant ($F_{1,7}=4.389$, $P<0.005$).

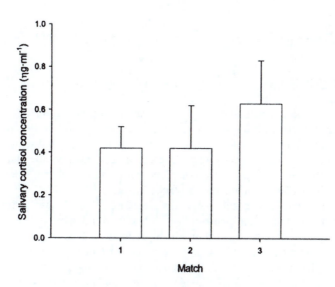

Figure 1. Mean ± SEM salivary cortisol concentration after each match during a period of 3 matches in 8 days.

The data for sIgA concentrations (Figure 2) after the matches showed a progressive decrease from 178.13 ± 55 mg·ml⁻¹ observed after the first match of the sequence to 148.68 ± 76 mg·ml⁻¹ and 141.25 ± 44 mg·ml⁻¹ after the second and third matches respectively. This decrease in the concentration of sIgA approached significance ($F_{2,8}=1.940$, $P=0.072$).

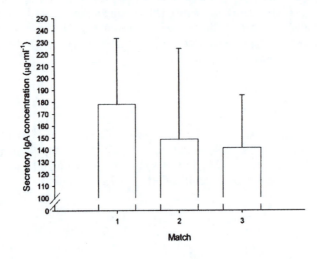

Figure 2: Mean ± SEM sIgA concentration after each match during a period of 3 matches in 8 days.

4. DISCUSSION

The data from the present study suggest that an intense period of fixtures, where 3 matches are played in 8 days, may result in an elevated concentration of salivary cortisol and a suppression of sIgA after the third match compared with the first one of the sequence. These changes were evident as trends in the data, with the likelihood that one (or two) of the subjects was able to tolerate the competitive schedule without alterations in the salivary measures.

Although in the present study the decrease in sIgA concentrations only approached statistical significance ($P=0.072$) the overall post-exercise effect size of 0.47 would indicate a medium-sized reduction between the games. This observation supports the findings that a single bout of high-intensity exercise is sufficiently demanding to reduce the concentration of sIgA post-exercise (Gleeson and Pyne, 2000; Engels *et al.*, 2004). In addition, long-term training has been demonstrated to reduce sIgA levels and have a detrimental impact on performance, during a 6-month training period; elite swimmers who were classified as stale had significantly lower sIgA levels than well-adapted swimmers (Mackinnon and Hooper, 1994). In contrast, Filaire *et al.* (2003) reported that sIgA levels remained unaltered throughout a competitive soccer season. However, the samples were collected at rest in the morning and the period between sample points was much greater than in the present study.

A reduction in the saliva concentration of IgA has been implicated as a possible causal factor in the increased susceptibility of athletes to upper respiratory tract infections or URTI (Tharp, 1991). Mackinnon *et al.* (1991) reported that more than 90% of athletes who developed URTI exhibited low sIgA concentrations prior to infection. Therefore a period of intense fixtures, which resulted in a reduced concentration of sIgA, may increase the likelihood of soccer players developing upper respiratory tract infections. One possible explanation for the increased susceptibility to URTI following exercises is the 'open window' theory (Nieman, 1994). The 'open window' is a period of time lasting 3-24 hours, where aspects of the immune system are temporally suppressed. During this window an infectious agent may be able to gain a foothold on the host and increase the risk of an opportunistic infection (Pedersen and Ullum, 1994). As a consequence, Neville *et al.* (2008) suggested that the regular monitoring of resting sIgA may be beneficial in terms of determining the risk of URTI and detecting overtraining in elite athletes.

The concentration of salivary cortisol remained unaltered after two matches and only exhibited an increase after the third match. This observation is similar to that of Filiaire *et al.* (2003) who reported a significant increase in cortisol concentration during the competitive soccer season. However, the mechanism for the increase remains unclear. One possible explanation for the increased cortisol response after the third match in the present study is that this event corresponded with the only victory in the sequence. Gonzalez-Bono *et al.* (1999) reported that cortisol concentrations increased in the victors (although not significantly) following a basketball match, suggesting a possible psychological influence on salivary cortisol. Furthermore, Haneishi *et al.* (2007) concluded that the cortisol response to a competitive game is influenced by both physiological and psychological variables.

In conclusion, the results from this study suggest that an intense period of fixtures may elevate the concentration of salivary cortisol and suppress sIgA. Falls in the saliva concentration of immunoglobulin A have been implicated as a possible causal factor in the increased susceptibility of athletes to upper respiratory tract infections. Therefore a period of intense fixtures may increase the likelihood of soccer players developing immunosuppression leading to upper respiratory tract infections. However, further investigation is required to substantiate the findings of the present study due to the sample size being limited to six subjects as a consequence of team selection and substitutions. Such investigation at a professional competitive level would require the co-operation of 2-3 teams to provide adequate subject numbers.

References

Akimoto, T., Nakahori, C., Aizawa, K., Kimura, F., Fukubayashi, T. and Kono, I., 2003, Acupuncture and responses of immunologic and endocrine markers during competition. *Medicine and Science in Sports and Exercise,* **35**, pp. 1296-1302.

Bishop, N.C., Blannin, A.K., Armstrong, E., Rickman, M. and Gleeson, M., 2000, Carbohydrate and fluid intake affect the saliva flow rate and IgA response to cycling. *Medicine and Science in Sports and Exercise,* **32**, pp. 2046-2051.

Chicharro, J.L., Lucia, A., Perez, M., Vaquero, A.F. and Urena, R., 1998, Saliva composition and exercise. *Sports Medicine,* **26**, pp. 17-27.

Ekstrand, J., Walden, M. and Hagglund, M., 2004, A congested football calendar and the wellbeing of players: correlation between match exposure of European footballers before the World Cup 2002 and their injuries and performances during that World Cup. *British Journal of Sports Medicine,* **38**, pp. 493-497.

Engels, H.J., Fahlman, M.M., Morgan, A.L. and Formolo, L.R., 2004, Mucosal IgA response to intense intermittent exercise in healthy male and female adults. *Journal of Exercise Physiology Online,* **7**, pp. 21.

Filaire, E., Bernain, X., Sagnol, M. and Lac, G., 2001a, Preliminary results on mood state, salivary testosterone:cortisol ratio and team performance in a professional soccer team. *European Journal of Applied Physiology,* **86**, pp. 179-184.

Filaire, E., Duche, P., Lac, G. and Robert, A., 1996, Saliva cortisol, physical exercise and training: Influences of swimming and handball on cortisol concentrations in women. *European Journal of Applied Physiology and Occupational Physiology,* **74**, pp. 274-278.

Filaire, E., Lac, G. and Pequignot, J.M., 2003, Biological, hormonal, and psychological parameters in professional soccer players throughout a competitive season. *Perceptual and Motor Skills,* **97**, pp. 1061-1072.

Filaire, E., Sagnol, M., Ferrand, C., Maso, F. and Lac, G., 2001b, Psychophysiological stress in judo athletes during competitions. *Journal of Sports Medicine and Physical Fitness,* **41**, pp. 263-268.

Gleeson, M., McDonald, W.A., Pyne, D.B., Clancy, R.L., Cripps, A.W., Francis, J.L. and Fricker, P.A., 2000, Immune status and respiratory illness for elite swimmers during a 12-week training cycle. *International Journal of Sports Medicine,* **21**, pp. 302-307.

Gleeson, M., McDonald, W.A., Pyne, D.B., Cripps, A.W., Francis, J.L., Fricker, P.A. and Clancy, R.L., 1999, Salivary IgA levels and infection risk in elite swimmers. *Medicine and Science in Sports and Exercise,* **31**, pp. 67-73.

Gleeson, M. and Pyne, D.B., 2000, Exercise effects on mucosal immunity. *Immunology and Cell Biology,* **78**, pp. 536-544.

Gonzalez-Bono, E., Salvador, A., Serrano, M.A. and Ricarte, J., 1999, Testosterone, cortisol, and mood in a sports team competition. *Hormones and Behavior,* **35**, pp. 55-62.

Haneishi, K., Fry, A.C., Moore, C.A., Schilling, B.K., Li, Y. and Fry, M.D., 2007, Cortisol and stress responses during a game and practice in female collegiate soccer players. *Journal of Strength and Conditioning Research,* **21**, pp.583-588.

Lamm, M.E., 1997, Interaction of antigens and antibodies at mucosal surfaces. *Annual Review of Microbiology,* **51**, pp. 311-340.

Mackinnon, L.T., Ginn, E. and Seymour, G.J., 1991, Temporal relationship between exercise-induced decreases in salivary IgA concentration and subsequent appearance of upper respiratory illness in elite athletes. *Medicine and Science in Sports and Exercise,* **23**, S45.

Mackinnon, L.T., Ginn, E. and Seymour, G.J., 1993, Decreased salivary immunoglobulin-A secretion rate after intense interval exercise in elite kayakers. *European Journal of Applied Physiology and Occupational*

Physiology, **67**, pp. 180-184.

Mackinnon, L.T. and Hooper, S., 1994, Mucosal (secretory) immune-system responses to exercise of varying intensity and during overtraining. *International Journal of Sports Medicine,* **15**, S179-S183.

Nakamura, D., Akimoto, T., Suzuki, S. and Kono, I., 2006, Daily changes of salivary secretory immunoglobulin A and appearance of upper respiratory symptoms during physical training. *Journal of Sports Medicine and Physical Fitness,* **46**, pp. 152-157.

Navazesh, M. and Christensen, C.M., 1982, A comparison of whole mouth resting and stimulated salivary measurement procedures. *Journal of Dental Research,* **61**, pp. 1158-1162.

Neville, V., Gleeson, M. and Follant, J.P., 2008, Salivary IgA as a risk factor for upper respiratory infections in elite professional athletes. *Medicine and Science in Sports and Exercise,* **40**, pp. 1228-1236.

Nieman, D.C., 1994, Exercise, upper respiratory tract infection, and the immune system. *Medicine and Science in Sports and Exercise,* **26**, pp. 128-139.

Nieman, D.C. and Nehlsen-Cannarella, S.L., 1991, The effects of acute and chronic exercise on immunoglobulins. *Sports Medicine,* **11**, pp. 183-201.

Pedersen, B.K., 1991, Influence of physical activity on the cellular immune-system - mechanisms of action. *International Journal of Sports Medicine,* **12**, S23-S29.

Pedersen, B.K. and Ullum, H., 1994, NK cell response to physical activity; possible mechanisms of action. *Medicine and Science in Sports and Exercise,* **26,** pp. 140-146.

Pedersen, B.K., Woods, J.A. and Nieman, D.C., 2001, Exercise-induced immune changes - an influence on metabolism? *Trends in Immunology,* **22**, pp. 473-475.

Sari-Sarraf, V., Reilly, T., Doran, D.A. and Atkinson, G., 2007, The effects of single and repeated bouts of soccer-specific exercise on salivary IgA. *Archives of Oral Biology,* **52**, pp. 526-532.

Sari-Sarraf, V., Reilly, T., Doran, D.A. and Atkinson, G., 2008, Effects of repeated bouts of soccer-specific intermittent exercise on salivary IgA. *International Journal of Sports Medicine,* **29**, pp. 366-371.

Seidman, D.S., Dolev, E., Deuster, P.A., Burstein, R., Arnon, R. and Epstein, Y., 1990, Androgenic response to long-term physical training in male subjects. *International Journal of Sports Medicine,* **11**, pp. 421-424.

Shephard, R.J., Rhind, S. and Shek, P.N., 1994, Exercise and the immune-system - natural-killer-cells, interleukins and related responses. *Sports Medicine,* **18**, pp. 340-369.

Tabata, I., Atomi, Y., Mutoh, Y. and Miyashita, M., 1990, Effect of physical training on the responses of serum adrenocorticotropic hormone during prolonged exhausting exercise. *European Journal of Applied Physiology and Occupational Physiology,* **61**, pp. 188-192.

Tharp, G.D., 1991, Basketball exercise and secretory immunoglobulin A. *European Journal of Applied Physiology,* **63**, pp. 312-314.

Walsh, N.P., Bishop, N.C., Blackwell, J., Wierzbicki, S.G. and Montague, J.C.,

2002, Salivary IgA response to prolonged exercise in a cold environment in trained cyclists. *Medicine and Science in Sports and Exercise,* **34**, pp. 1632-1637.

Weicker, H. and Werle, E., 1991, Interaction between hormones and the immune-system. *International Journal of Sports Medicine,* **12**, S30-S37.

CHAPTER FIFTEEN

The relationship between performance on the 15-30 and other soccer-specific protocols and work-rate in match-play

J.M. Svensson, B. Drust and T. Reilly

Research Institute for Sport and Exercise Sciences, Liverpool John Moores University, Liverpool, UK

1. INTRODUCTION

Sports scientists operating in support roles in professional soccer clubs require tests of fitness that represent the physiological status of their players and are valid for the sport. In order to be useful in the regular monitoring of fitness and detecting small changes due to training interventions and recovery after periods of strenuous training or congested competitive fixtures, field tests have convenience for ease of administration to whole squads. Once their validity and utility have been demonstrated, the reliance on time-consuming laboratory tests is reduced and the field tests generate performance-related criteria that have immediate meaning for practice.

The most prominent tests of physiological function have been the maximal oxygen uptake ($\dot{V}O_{2\,max}$) for assessment of aerobic power and the lactate threshold test (T_{lac}) for assessment of aerobic capacity. Both variables have been related to performance in individual endurance events such as cycling and marathon racing (Davison et al., 2008). They have also been significantly correlated with work-rate in soccer match-play, expressed as distance covered in a game (Reilly and Thomas, 1976; Reilly, 1997). The specificity of the assessment protocol for soccer may be criticised, comprising a progressive run to voluntary exhaustion rather than the incorporation of frequent high-intensity efforts as occur in competitive soccer. The same comment may be levelled at field tests such as the progressive 20-metre shuttle run (Ramsbottom et al., 1988) designed to predict $\dot{V}O_{2\,max}$ in groups of athletes at one time.

The Yo-Yo tests designed by Bangsbo (1993) overcome this difficulty by including intermittent exercise within the test protocols. Performance on the tests has been shown to reflect aerobic capabilities and improve the correlation with aspects of performance in a match, notably the distance covered in high-intensity components of the overall work-rate (Krustrup et al., 2003). An alternative is the "15-30 Protocol" first proposed by Svensson and Drust (2005) and subsequently shown to detect the effects of an interval-training intervention (Svensson et al., 2009). Its advantages are that it includes both sub-maximal and maximal parts, so that the sub-maximal component may be used regularly by inclusion in the normal training sessions. The physiological responses are recorded for assessment purposes, whereas the performance component is represented by the duration of exercise before exhaustion arises. The duration of the high-intensity bouts of exercise is varied systematically so as to improve the correspondence with activity in matches. In this way the soccer-specific speed-endurance of individual players can be evaluated.

We compared performance on the 15-30 protocol to submaximal and maximal assessments of aerobic capabilities. The lactate threshold and maximal oxygen uptake were chosen for laboratory assessments, whereas the 20-metre shuttle run and the Yo-Yo Intermittent Recovery test were significantly correlated with physical performance as reflected in the total distance covered in a game among young professional players.

2. METHODS

2.1 Participants

Nineteen young professional players from a Coca-Cola Championship League One club in England were recruited. During the period of investigation, three players were released from the club and were excluded from the analysis. The characteristics of the remaining 16 players were: age, 17.1 ± 0.8 years; height 1.764 ± 0.065 m; body mass 72.1 ± 6.8 kg. No strenuous training sessions took place on the day preceding any of the test sessions. The players were informed of the benefits, risks and stresses of the assessments before providing written informed consent. The project was approved by the Research Ethics Committee of Liverpool John Moores University.

2.2 Field tests

The field tests were conducted on separate days in an indoor sports hall at the club's training premises. The tests were conducted according to the standard procedures of Ramsbottom *et al*. (1988) and Krustrup *et al*. (2003) for the 20-m shuttle run and the Yo-Yo Intermittent Recovery Test (YYIRT), respectively. The protocol for the "15-30" was that used by Svensson *et al*. (2009) when establishing its reliability.

Briefly, the 15-30 protocol consisted of a sub-maximal stage and a maximal stage commencing 30 s immediately following. The sub-maximal part is divided into five blocks, each containing seven cycles of short (2 x 15 m) and long (30 m) running bouts at a pace dictated by an audio-tape. A passive recovery of 6 s and 12 s follows the short and longer runs, respectively, both runs having to be completed within 6 s. The total distance covered in the five blocks is 2030 m. For the maximal stage (15-30$_{max}$), each 58-m bout contains two turns round cone markers set at different distances apart with a passive rest of 6 s between runs. The running bouts are repeated until the participant desists or can no longer complete the runs within the designated period of 6 s.

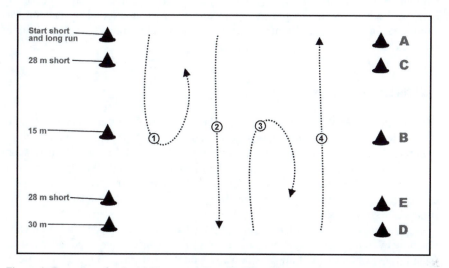

Figure 1. Test course for the 15-30 protocol (Part 2). The subjects started at A where they ran to B, turned in line with the cones and then ran back to C. This run had to be completed within 6 s. After the run a 6-s active rest period then followed where the subjects walked to A ready to start the next part of the run. They then ran from A to D within 6 s. After the run a 6-s passive rest period followed. These two runs made up one exercise bout. The total distance covered for one bout was 58 m. The subjects then started the next bout at D where they ran to B, turned and ran back to E, followed by an active 6-s rest period. In the rest period, the subjects walked to D ready to start the next run. The subjects then ran from D to A in 6-s followed by another 6-s rest period. The subjects repeated these exercise bouts until they could no longer complete the runs within the allocated time period.

The physiological responses included heart rate determined by means of short-range radio telemetry (Polar Electro Oy, Kempele). Data were recorded every 5 s and the overall mean value calculated. Blood samples from a fingertip were obtained at the end of each running bout and analysed for lactate concentration using the Lactate Pro (Arkray, Kyoto) machine. Perceived exertion was rated at the same time using Borg's (1998) scale of 6-20.

2.3 Laboratory assessments

Lactate threshold was determined as the velocity corresponding to a blood lactate concentration of 2 mM and the onset of blood lactate accumulation at V-4 mM. The test was performed on a motorised treadmill (h/p Cosmos Pulsar, Nüssdorf-Traunstein). After an initial warm-up for 5 min at 8 km.h^{-1}, the test speeds commenced at 10 km.h^{-1} with a step-wise increase of 1 km.h^{-1} every 4 min. After each stage there was a rest of 30 s during which the player rested with feet to the side of the treadmill belt to facilitate blood sampling.

Maximal oxygen uptake was determined using a graded exercise test to voluntary exhaustion according to Winter *et al.* (2008). The test was commenced after the establishment of V-4 mM. Oxygen uptake was recorded continuously with an on-line breath-by-breath gas analysis system (Metamax Cortex Biophysik, Leipzig) connected to the player by means of a facemask. The system was calibrated for each player by means of a 3-l syringe and known concentrations of O_2 and CO_2.

2.4 Work-rate analysis

Sixteen matches were filmed during the 2005-2006 competitive season to determine activity patterns and work-rates during matches. Youth games in the under-19 North West Youth Alliance League, F.A. Youth Cup and reserve team matches from the Pontins League West were included, during which the team used a 4-4-2 outfield formation. Only outfield players completing the full game were considered, reducing the sample size to 13 for the correlation analysis with the fitness assessment data.

The games were recorded using a digital video camera placed at an elevated position over the half-way line. Five motion categories were defined as static (standing), walking (including sideways and backwards), jogging (including sideways and backwards), cruising and sprinting. The frequencies of jumping, periods on the ground, jockeying for position were also noted. The data were then collated for distance covered in a game and in each of the movement categories as well as placed on a time base.

The method was subjected to assessment of its intra-observer reliability (repeatability) and inter-observer variation (objectivity) by re-analysing a 45-min sample of four matches. The strength of exact agreement for the motion categories was deemed very good for both repeatability and objectivity according to the kappa statistic (Altman, 1991). The coefficient of variation, limits of agreement and coefficient of variation for the limits of agreement (Bland and Altman, 1986) were calculated for the total distances covered. The 95% limits of agreement in all assessments fell within 1% of the distance covered and the co-efficient of variation of the limits of agreement ranged from 3.1 to 5.7%. Based on these results and the low systematic bias for both repeatability and objectivity, the method was confirmed as acceptable.

2.5 Statistical analysis

The data are presented as means ± S.D. Correlations between fitness measures and work-rates were explored using Pearson's Product Moment coefficients. Differences between playing positions were examined where appropriate using one-way analysis of variance (ANOVA). Comparisons of peak physiological measures in the various tests and between times spent in different motion categories were treated with paired t-tests. Wilcoxon's signed rank test was used where parametric assumptions were not fulfilled.

3. RESULTS

3.1 The 15-30 protocol

The heart rate averaged 166 ± 8 beats.min^{-1} during Part 1 (15-30$_{submax}$) and 186 beats.min^{-1} for the maximal stage. The heart rate increased progressively with each of the five blocks of 15-30$_{submax}$ ($F_{5, 48} = 8.06$; $P < 0.05$), but no difference was noted for the recovery periods. A similar increase was observed in perceived exertion ($F= 25.98$; $P < 0.001$). The distance covered in 15-30$_{max}$ averaged 1327 ± 677 m, varying significantly with playing position (goalkeepers n=2; 580 ± 246), but not among the outfield players: defenders n=5; 1264 ± 733 m; forwards n=4; 1189 ± 121 m; midfielders n=5; 1798 ± 758 ($F=1.28$; $P > 0.05$).

3.2 The 20-m shuttle run

During the test the distance covered during the test amounted to 2271 ± 309 m. The $\dot{V}O_{2\ max}$ estimated from the terminal stage was 53.08 ± 4.1 ml.kg^{-1}.min^{-1}. The peak heart rate was 199 ± 8 beats.min^{-1}.

YYIRT tests: the total distance covered in the test was 1857 ± 434 m. The peak heart rate at the end of the test was 194 ± 9 beats.min^{-1}. At this time the rating of perceived exertion reached 19.1 ± 1.7.

3.3 Laboratory assessments

The V-2 mM value used to determine T_{lac} corresponded to a running speed of 12.8 ± 1.1 km.h^{-1}. This value represented $73.8 \pm 4.4\%$ $\dot{V}O_{2\ max}$ and $84.4 \pm 3.9\%$ peak heart rate. The second measure of V-4 mM was found at 14.1 ± 1.0 km.h^{-1}, constituting a relative loading of $83.2 \pm 5.2\%$ $\dot{V}O_{2\ max}$, and $89.8 \pm 2.7\%$ of peak heart rate. The mean value of $\dot{V}O_{2\ max}$, 62.4 ± 5.0 ml.kg^{-1}.min^{-1}, was significantly higher than that estimated from performance in the 20-m shuttle run ($P < 0.01$).

3.4 Work-rate analysis

The mean distances covered for the players in matches were 4950 ± 479 m in the first half and 4791 ± 577 m for the second half, demonstrating a non-significant decrease of 3% ($P > 0.05$). Similarly a fall in high-intensity exercise categories of 14% from 766 ± 254 to 662 ± 265 m did not reach statistical significance ($P > 0.05$). The distance varied between playing positions, being lowest in defenders (9328 ± 541 m) and highest in midfielders (10235 ± 1088 m) with intermediate values observed in the forwards (9768 ± 1299 m).

Table 1. Laboratory test results for the professional youth scholars compared to playing position.

Playing position	Running speed $(km.h^{-1})$ at T_{lac}	Running speed $(km.h^{-1})$ at V-4 mM	$\dot{V}_{O_2\,max}$ $(l.min^{-1})$	$\dot{V}_{O_2\,max}$ $(ml.kg^{-1}.min^{-1})$	$\dot{V}_{O_2\,max}$ $(ml.kg^{-0.75}.min^{-1})$	HR_{peak} $(beats.min^{-1})$	Time to exhaustion (s)
Goalkeepers (N=2)	11.5 ± 0.7	13.6 ± 0.7	4.03 ± 0.2	56.31 ± 1.0	169.5 ± 1.0	199 ± 4	160.5 ± 30.4
Defenders (N=5)	13.4 ± 0.9	14.5 ± 0.6	4.43 ± 0.4	63.03 ± 5.8	182.0 ± 17.0	199 ± 15	200.4 ± 45.3
Midfielders (N=5)	13.0 ± 1.4	14.5 ± 1.7	4.42 ± 0.6	63.61 ± 4.7	181.2 ± 18.2	200 ± 10	228.2 ± 28.8
Forwards (N=4)	12.3 ± 0.5	13.7 ± 0.8	4.36 ± 0.4	60.33 ± 2.9	177.1 ± 7.9	189 ± 8	149.8 ± 16.4 *
Mean $\pm$ SD (N=16)	12.8 ± 1.1	14.3 ± 1.1	4.36 ± 0.4	62.41 ± 4.8	179.0 ± 14	197 ± 11	199.5 ± 46.1

* = significant (P<0.05) to midfielders

3.5 Correlation analysis

Relative heart rate for $15\text{-}30_{submax}$ was significantly correlated with distance covered in $15\text{-}30_{max}$ ($r = -0.561$, $P < 0.05$). Correlation coefficients for heart rate on the submaximal part of the test did not reach statistical significance for $\dot{V}O_{2\,max}$ (either relative or scaled: $r = -0.332$ and $r = -0.310$, both $P > 0.05$), V-4 mM ($r = -0.124$; $P > 0.05$), the 20-m shuttle run ($r = -0.416$; $P > 0.05$) or distance covered in YIRT ($r = -0.096$; $P > 0.05$). There was no significant correlation for distance covered overall or in high-intensity activity during match-play ($r = -0.139$ and -0.27; both $P > 0.05$).

The distance covered in completing the $15\text{-}30_{max}$ test was highly correlated with relative $\dot{V}O_{2\,max}$ ($r = 0.831$; $P < 0.01$), scaled $\dot{V}O_{2\,max}$ ($r = 0.791$; $P < 0.05$), running speed at V-4 mM ($r = 0.600$; $P < 0.05$) and distance covered on the 20-m shuttle run ($r = 0.745$, $P < 0.05$) for all 16 players. The correlations with the distance covered in YIRT, and high-intensity distance (cruising and sprinting) during match-play were significant ($r = 0.448$ and 0.414; both $P < 0.05$), but not for the distance in match-play. A moderate correlation ($r = 0.331$, $P > 0.05$) was noted between the distances covered in the $15\text{-}30_{max}$ test and distance covered in match play - and between the distances covered in the YIRT ($r = 0.448$, $P > 0.05$) and 20-m shuttle run tests ($r = 0.745$, $P < 0.05$).

4. DISCUSSION

The most important finding in the current study was the partitioning of the physiological components in the $15\text{-}30_{max}$ test into its aerobic characteristics. Based on the correlation analysis, 69% of the overall variance in test performance could be explained by maximal aerobic power ($ml.kg^{-1}.min^{-1}$) and 36% by aerobic capacity as reflected in V- 4 mM or lactate threshold. The remaining non-overlapping variance could be attributed to soccer-specific factors and random error.

Both parts of the test had practical value, the heart rate response in $15\text{-}30_{submax}$ being moderately predictive of distance covered in $15\text{-}30_{max}$. This test has previously been shown to be sensitive to aerobic interval training and to have high reliability (Svensson *et al.*, 2009). It may be more useful as a tool to identify transient "underperformance" or fatigue in individual players than in predicting physical performance in match-play. It seems that apart from blood lactate responses to fixed sub-maximal intensities, maximal loading is required for a more complete assessment of performance capability in soccer.

The correlation analysis would suggest that the global aspects of endurance are common to both the 20-m shuttle run and the $15\text{-}30_{max}$ tests. The under-estimation of $\dot{V}O_{2\,max}$ in the 20-m shuttle run underlines the difficulty faced by the need to solicit a complete commitment of players to volitional exhaustion. In contrast, the same peak heart rates were reached in the YIRT and $15\text{-}30_{max}$ tests, indicating a maximal effort. The main difference between these two tests which purport to incorporate the high-intensity exercise bouts akin to soccer play, may lie in their anaerobic endurance components. The YIRT test emphasises the ability to recover quickly whereas the 15-30$_{max}$ tests probe the ability to maintain performance under conditions of brief recoveries, thus placing a greater dependence on oxygen transport and utilisation.

The failure to identify clear predictors of work-rate (except for the high-intensity components of the activity patterns) in matches merits explanation and a parsimonious mechanistic link between fitness measures and work-rate variables may be an unrealistic expectation at the youth level of play. Activity patterns may be unstable as players undergo development and many of the players may not have their capabilities taxed to the full in all matches. Game-to-game variations, different competitions and level of opposition could also affect the results. Nevertheless, there were indications that fitness measures are relevant to performance as reflected in the work-rate profiles, based on the correspondence of the superiority of the midfield players and the lower ranking of defenders on both work-rate and fitness measures.

In conclusion, the current study supports the continued use of the 15-30 test for soccer, the submaximal part for prophylactic purposes and the maximal part for performance capability. Despite the domination of its aerobic power contribution, the maximal test offers a simultaneous assessment of anaerobic endurance integrated into the test result. Based on the similarities and differences with other field tests and the lack of a deterministic link to work-rate in matches, it is doubtful if the complex activity of soccer can be simulated in one single assessment protocol.

REFERENCES

Altman, D., 1991, *Practical Statistics for Medical Research* (London: Chapman and Hall).
Bangsbo, J., 1993, *Yo-Yo Testene* (Brondby, Denmark: Denmark Idraetsforbund).
Bland, J.M. and Altman, D.G., 1986, Statistical methods for assessing agreement between two methods of clinical measurement. *Lancet*, **1**, pp. 307-310.
Borg, G., 1998, *Borg's Perceived Exertion and Pain Scales* (Champaign, Ill; Human Kinetics).
Davison, R., Jobson, S., de Koenig, J. and Balmer, J., 2008, The science of time-trial cycling. In *Science and Sports: Bridging the Gap*, edited by Reilly, T. (Shaker: Maastricht), pp. 77-93.
Krustrup, P., Mohr, M., Armstrup, T., Rysgaard, J., Johansen, J., Steensberg, A., Pedersen, P.K. and Bangsbo, J., 2003, The Yo-Yo intermittent recovery test: physiological response, reliability and validity. *Medicine and Science in Sports and Exercise*, **35**, pp. 695-705.
Ramsbottom R., Brewer, J. and Williams, C., 1988, A progressive shuttle run test to estimate maximal oxygen uptake. *British Journal of Sports Medicine*, **22**, pp. 141-144.
Reilly, T., 1997, Energetics of high-intensity exercise (soccer) with special

reference to fatigue. *Journal of Sports Sciences*, **15**, pp. 257-263.

Reilly, T. and Thomas, V., 1976, A motion analysis of work rate in different positional roles in professional soccer match-play. *Journal of Human Movements Studies*, **2**, pp. 87-97.

Svensson, M. and Drust, B., 2005, Testing soccer players. *Journal of Sports Sciences*, **23**, pp. 601-618.

Svensson, M., Conway, P., Drust, B. and Reilly, T., 2009, Performance on two soccer-specific high-intensity intermittent running protocols. In *Science and Football VI*, edited by Reilly, T. and Korkusuz, F. (London: Routledge), pp. 350-356.

Winter, E.M., Jones, A.M., Dawson, R.C.R, Bromley, P.D. and Mercer, T.M., 2008, Sport and Physiology Testing: Guidelines: The British Association of Sport and Exercise Sciences Guide, *Volume 1 Sport Testing* (London: Routledge).

CHAPTER SIXTEEN

Directional change, multifactorial analysis of movement

A. Ravaschio[1], R. Sassi[2], A. Tibaudi[2] and N.A. Maffiuletti[3]

[1]Laboratory of Bioengineering, Osp. San Francesco, Barga, Italy
[2]Professional Athletic Trainers, Italy
[3]Neuromuscular Research Laboratory, Schulthess Clinic, Zurich, Switzerland

1. INTRODUCTION

One of the most discussed aspects amongst athletic coaches engaged in soccer training is related to the improvement of specific strength for the sport (Hoff and Hegelrud, 2004; Bangsbo *et al.*, 2006). In recent years numerous methods that require the use of equipment and other apparatus, whose utility is still debated, have been proposed and applied.

Moreover, one of the most important problems that technical teams must deal with is illustrated by the methods employed for the prevention of accidents. Recently it has been shown that most accidents in professional soccer are related to lower limbs, and that one-fourth of all knee injuries are not connected to direct contact with an opponent (Junge and Dvorak, 2004). Moreover, it has been shown that the absence of footballers due to accidental injury has a negative financial effect on the club. Running with directional changes (CdD) is one of the main ways in which players move during the game. Such a motion is repeated frequently during the game and includes phases of concentric type (e.g., accelerations) and eccentric type (e.g., decelerations) work performed at different speeds. However, a full analysis of the forces produced during a CdD at the ankle and knee joints (i.e., the joints which are most likely to be injured) has not been realized up to now. The present study was aimed to quantify the forces applied to the ankle and the knee joints during a 180° CdD (shuttle type) and afterwards to compare them with a vertical counter-movement jump (CMJ); the latter exercise is usually utilized in soccer training for improvement of "explosive strength".

2. METHODS

Five professional footballers playing in the championship of the Italian C2 series (age: 23 ± 4 years; height: 1.75 ± 0.06 m; body mass: 72 ± 7 kg; mean ± SD) performed a series of 180° CdD (stationary start and 5 metres acceleration phase) at the highest intensity and a CMJ series. For movement analysis the Vicon 612 system (Oxford Metrics Ltd., Oxford, UK) consisting of two force plates (Kistler Instrumente AG, Winterthur, Switzerland) with six components, was utilized. For calculating torque values at the joints, the Davis "Plug and Gait" model designed for "gait analysis" was utilized. The eccentric (negative power) and concentric (positive power) phases of the movement in the three planes (flexion/extension, abduction/adduction and rotation) were distinguished from the power output diagram.

3. RESULTS and DISCUSSION

Maximal values of eccentric and concentric torque expressed in the three planes (frontal, sagittal and horizontal) for the knee joint (Table 1) and ankle joint (data not reported) were higher during CdD compared to CMJ.

Table 1. Torques measured at the knee joint during a CdD and a CMJ. Mean values ± SD.

	Eccentric torque (Nm)			Concentric torque (Nm)		
	Flex/Extension	Abd/Adduction	Rotation	Flex/Extension	Abd/Adduction	Rotation
CdD	1512.7 ± 58.2	1137.9 ± 240.1	345.7 ± 16.8	1107.3 ± 64.8	1463.8 ± 600.6	308.0 ± 146.5
CMJ	1171.7 ± 319.5	454.2 ± 238.1	127.7 ± 39.1	1042.6 ± 484.3	563.6 ± 328.9	139.1 ± 42.3

In the period of training it is important to subject the joints to elevated loadings in order to secure adaptations to soccer; however, actions and movements must be similar to the actions the player will be making during the game, respecting the specificity of training principle. We have also found that increasing squat loading with 50% overload of the subjects' body weight, caused flexor-extensor torques at the knee joint to increase less than 10% compared to that in CMJ, while adduction-abduction torques remained unchanged. These observations reflect the non-specific part of the action.

4. CONCLUSIONS

The results of this study have demonstrated that in 180° CdD the physical loadings are higher compared to the exercises classically utilized for strength improvement in soccer (CMJ), even if the type of applied force is different. The present research confirms the necessity of utilizing, in professional and non-professional soccer players, CdD exercises organized in series and repetitions, since such activity generates high loadings in accordance with soccer-specific motor actions. The CdD may be considered "training" exercises like other movements and, if performed systematically they lead to improving the forces expressed by the footballer in a movement similar to that occurring in the game. Because of achieving a better coordination, their use during training represents a preventive measure against joint lesions connected to indirect accidental injury.

References

Bangsbo, J., Mohr, M. and Krustrup, P., 2006, Physical and metabolic demands of training and match play in the elite football player. *Journal of Sports Sciences*, **24**, pp. 665-674.

Hoff, J. and Hegelrud, J., 2004, Endurance and strength training for soccer players: physiological considerations. *Sports Medicine*, **34**, pp. 165-180

Junge, A. and Dvorak J., 2004, Soccer injuries: a review of incidence and prevention. *Sports Medicine*, **34**, pp. 929-938.

CHAPTER SEVENTEEN

Video analysis of football injuries at the Asian Cup 2007

N. Rahnama[1] and M. Zareei[2]

[1]University of Isfahan, Isfahan, Iran
[2]University of Tehran, Tehran, Iran

1. INTRODUCTION

Soccer is one of the most popular spectator sports in the world. About 250 million licensed players in 204 countries are registered with FIFA, and about 1% participate at the professional level (Andersen *et al.*, 2004a). The risk of injury in this sport is also considerable. The overall level of injury for a professional player has been reported to be about 1000 times higher than that of a high risk industrial worker (Hawkins and Fuller, 1999). Injuries to soccer players not only threaten their health but also can be an economic burden to the individual and entail social and medical costs (Yoon et *al.*, 2004).

To minimize the number of injuries and the associated costs, avoid the early retirement of professional soccer players, and provide a safe and healthy sports experience, preventative programmes are required. In order to suggest preventative strategies specific to soccer, it is necessary to have detailed information on the injury mechanisms involved. It is difficult to gain this information on the basis of feedback from injured players because of recall bias (Andersen *et al.*, 2004b). As most elite soccer matches are televised, the use of video recordings instead of player interviews may improve our ability to identify and understand the injury mechanisms and other factors correlated with injury.

The incidence of injury varies depending on the type of tournament and the characteristics of the players and seems to be significantly higher in men than in women playing the same type of tournament (Junge *et al.*, 2004b). The incidence of injury in men's international soccer tournaments ranges from 51 to 144 injuries per 1000 match hours, equivalent to approximately 2 to 3 injuries per match (Junge *et al.*, 2004a). For example, Yoon *et al.* (2004) reported 140 injuries/1000 playing hours in the 2000 Asian Cup and Junge *et al.* (2004b) described for some male soccer World Cup tournaments an incidence of 114 injuries/1000 playing hours. These authors (Junge *et al.*, 2004a) reported 81 injuries/1000 playing hours for the 2002 World Cup finals and Dvorak *et al.* (2007) found 69 injuries/1000 playing hours in the 2006 World Cup. Most researchers reported the majority of soccer injuries are caused by player to player contact (Junge *et al.*, 2004 a, b; Yoon *et al.*, 2004).

Although there have been many studies conducted in Europe on the incidence of injuries during soccer tournaments, few studies have been carried out in Asia.

Thus the aim of this study was to analyze the incidence, circumstances, and characteristics of injury during the Asian Cup 2007 by using a video analysis system.

2. MATERIALS and METHODS

Thirty-one out of thirty-two matches of this tournament, played over a period of 22 days, were recorded by using a video recorder and analyzed using a computer and video (Samsung, SV-A140G). The video was paused after every injury that occurred. In this study a player was defined as injured if he required any form of treatment from the medical staff during the match (Rahnama *et al.*, 2002; Yoon *et al.*, 2004). The time of injury, zone of the pitch, stage of play, mechanism of the injury and anatomical location of the injury were notated on a specifically designed record sheet. The positional role of the player as defined at www.afcasiancup.com was also noted. The six positions corresponded to goalkeeper, mid-defender, line defenders, centre midfielders, line midfielder and forwards (in accordance with Andersen *et al.*, 2004a). The playing time was divided into six periods of 15 minutes. At the knock-out stage, if the game finished even, two extra time periods of 15-min each were included. The pitch in respect of direction of play was divided into 18 zones (Rahnama *et al.*, 2002) as shown in Figure 1. The number of hours that players were exposed to the risk of trauma was calculated by assuming that 22 players were playing each match and that each match lasted for 100 minutes (45 minutes' standard play plus 5 minutes' added time for each half) (Hawkins and Fuller, 1996; Morgan and Oberlander, 2001; Junge *et al.*, 2004b;). Injury frequency rates (IFRs) were calculated as the number of injuries per 1000 hours of competition.

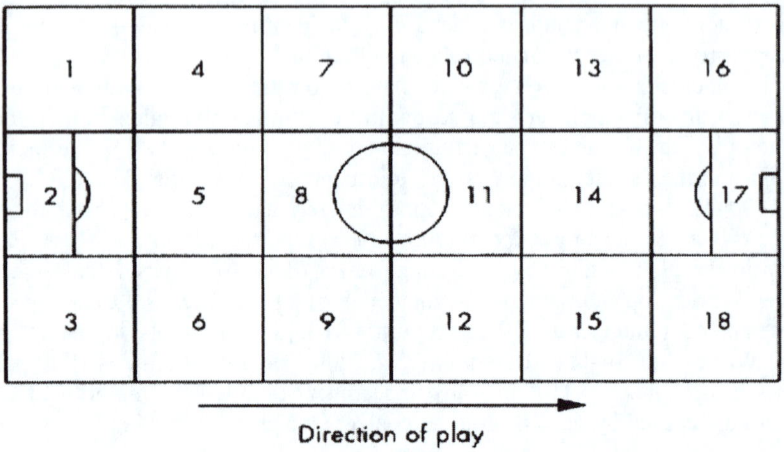

Direction of play

Figure 1. The zones of pitch demarcated for analysis of injuries events.

The data were analyzed using χ2 to compare categories, and a level of P<0.05 was used to indicate significance. Also Poisson Distribution (Z test) was used for comparison of injury incidence between knockout stages and first-round stages.

3. RESULTS

Altogether, 125 injuries were recorded for the 31 matches of the Asian Cup 2007. This figure represents an overall injury rate of 110 per 1000 match hours or 4 injuries per match.

Figure 2 shows an analysis of the match injuries with respect to their time of occurrence. Significantly (P<0.01) more injuries were observed during the final 15 minutes of the first half and the final 15 minutes of the second half. There were also significantly (P<0.01) more injuries recorded in the second half compared with the first half of matches (54.4% v 39.2%); 6.4% of the injuries occurred during added time.

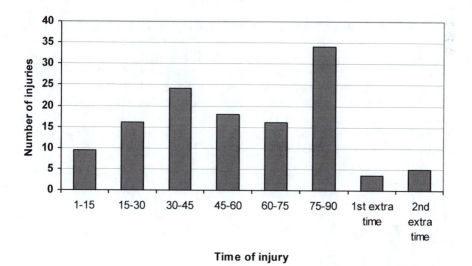

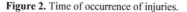

Figure 2. Time of occurrence of injuries.

There was a significant difference (*P* = 0.001) in the number of injuries between the zones of pitch. The most injuries occurred in the goal areas (2, 17) while 46.6% of injuries occurred in the defending area, 28.8% in the midfield area, and 24.6% in the attacking area (Figure 3).

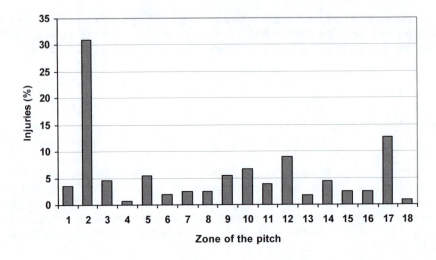

Figure 3. Percentage of injury per zone.

In the present study, the players experienced higher incidences of injury in the knockout stage than in the first-round matches. During the knockout stages, there were 4.5 injuries per game, and 3.8 injuries per game were noted during the first-round stages, although this difference was not significant ($Z = 0.35$, $P > 0.05$). Figure 4 shows the injury mechanisms that could be identified. The rate of injuries caused by contact between players was much higher than in non-contact circumstances ($P < 0.05$). More than three-quarters of the injuries were preceded by contact with an opponent. The commonest causes of injury involved aerial challenges (28%) and tackling duels (34%).

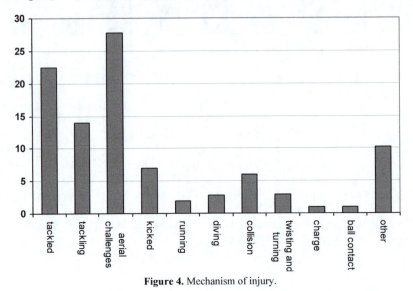

Figure 4. Mechanism of injury.

The majority of injuries affected the lower limb (56%), with 20% involving the lower leg and 14.4% involving the ankle. The injuries according to anatomical location are shown in Figure 5.

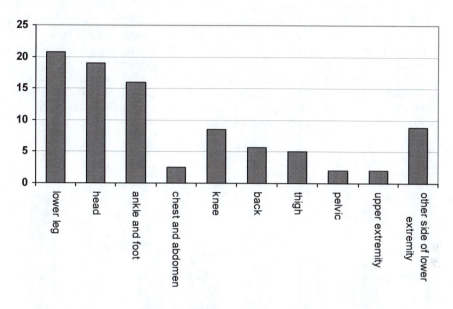

Figure 5. Anatomical location of injuries.

During the period of this study, goalkeeper was the most frequently injured playing position (P = 0.003). The forwards, line defenders, centre midfielders, mid-defenders and line midfielders respectively had lower injury rates (Table 1).

Table 1. Injury rate by player position.

Position	Number of injuries	Injury rate (per 1000 hours)
Defender Line	27	130.6
Mid-Defender	16	77.4
Centre Midfielder	24	116.1
Line Midfielder	11	53.2
Forward	29	140.3
Goalkeeper	18	174.2
Total	125	109.7

4. DISCUSSION

The purpose of this study was to document the incidences and patterns of injuries among elite Asian footballers. The definition of soccer injuries still remains controversial. In this study, a player was defined as injured if he required any form of treatment from the medical staff during the match (Hawkins and Fuller, 1996; Rahnama *et al.*, 2002, Yoon *et al.*, 2004). Such definition tends to increase the incidence of injuries compared with medical records (Yoon *et al.*, 2004).

This incidence of injury during the Asian Cup 2007 was higher than that recorded for the World Cup competitions of 1998, 2002, 2006 and the Olympic Games 2004. However, the figure is lower than that recorded for the Asian Cup 2000 and the Olympic Games (see Figure 6).

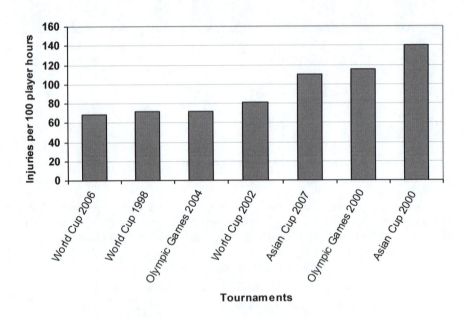

Figure 6. Injury frequency rates in different international tournaments.

The higher incidences of injury in this study may reflect the lower skill levels of Asian football players. They might also be linked to their different physical characteristics (Yoon *et al.*, 2004).

Consideration of the time during a match at which injury occurs highlights the final 15 minutes as the period of greatest risk of injury in each half (Figure 1). The significantly higher level of injury found in the second half of matches supports the results obtained by Hawkins and Fuller (1999) and Junge *et al.* (2004b). Others have reported that more injuries occur in the first half of matches (Junge *et al.*, 2004a). Muscle fatigue has previously been identified as a factor in injury causation (Hawkins and Fuller 1999), and onset of fatigue could partly explain the greater risk of injury observed during the final 15 minutes of each half.

There was a significant relationship between the frequency of injury and the zone of pitch. The most injuries occurred in the goal area (zones 2 and 17). Rahnama *et al.* (2002) reported more actions with injury potential occurred in the goal areas (zones 2 and 17). This pattern probably reflects the intense actions occurring as a result of efforts to score goals by forward players and to protect the goal area by opponents.

In this study, more injuries were observed in the knockout stages rather than in the first-round matches. Players in the knockout matches might have spent more time playing and training than players in the first round. Accumulated fatigue could therefore be a risk factor in injury during the knockout stages. This cumulative fatigue could be a crucial factor in centralized tournaments, where all matches are crucial and sometimes played within 2 days (Yoon *et al.,* 2004).

Player-to-player contact injury mechanisms, including aerial challenge, tackling, being tackled, and collisions, accounted for 78% of all injuries recorded in this study. Similar levels have been previously reported by others (McMaster and Walter, 1978; Junge *et al.,* 2004a, b; Yoon *et al.,* 2004).

The majority of soccer-related injuries affected the lower limb (56%). Demands of the game include a great deal of running, kicking, jumping, and contact with opposing players that could contribute to the incidence of injury (Rahnama *et al.,* 2002). Others have reported broadly similar results (McMaster and Walter, 1978; Junge *et al.,* 2004a; Price *et al.,* 2004; Yoon *et al.,* 2004; Junge *et al.,* 2006; Dvorak *et al.,* 2007) whereas McMaster and Walter (1978) did not report injuries to the lower limbs. In the extensive review by Inklaar (1994), the lowest percentage of lower limb injuries was reported as 61%, a figure higher than that observed in the current data. The upper leg (20%) was the most injured body part which may have been because of the large muscle mass and the large area exposed (Wong and Hong, 2005).

In the present study, goalkeepers sustained more injuries than players in other positional roles. Ostojic (2003) reported similar results. In addition, goalkeepers sustained injury more frequently at specific anatomical locations (head and hand) than other positional roles. This is probably due to goalkeepers' specific positional tasks and rules allowing the use of hands during the game.

Video analysis provides a convenient tool for investigating injuries occurring during televised tournaments, although it has limitations. Incidences off-screen may be missed, for example. Hawkins and Fuller (1996) estimated that at least 54 injuries were missed when using video footage to analyse the cause of injuries during the 1994 World Cup. In future studies, inter-observer and intra-observer reliability assessments are recommended to quantify the degree of error attributable to human factors.

References

Andersen, T.E., Tenga, A., Engebretsen, L. and Bahr, R., 2004a, Video analysis of injuries and incidents in Norwegian professional football. *British Journal of Sports Medicine*, **38**, pp. 626–631.

Andersen, T.E., Arnason, A. and Engebretsen, L., 2004b, Mechanisms of head injuries in elite football. *British Journal of Sports Medicine*, **38**, pp. 690-696.

Dvorak, J., Junge, A., Grimm, K. and Kirkendall, D., 2007. Medical report from the 2006 FIFA World Cup, Germany. *British Journal of Sports Medicine*, **41**, pp. 578-581.

Hawkins, R.D. and Fuller, C.W., 1999. A prospective epidemiological study of injuries in four English professional clubs. *British Journal of Sports Medicine*, **33**, pp.196-203.

Hawkins, R.D. and Fuller, C.W., 1996, Risk assessment in professional football: an examination of accidents and incidents in the 1994 World Cup finals. *British Journal of Sports Medicine*, **30**, pp.165-70.

Inklaar, H., 1994, Soccer injuries. I: Incidence and severity. *Sports Medicine*, **18**, pp. 55-73.

Junge, A., Dvorak, J. and Graf-Baumann, T., 2004a, Football injuries during the World Cup 2002. *American Journal of Sports Medicine*, **32**, S23-27.

Junge, A., Dvorak, J. and Graf-Baumann, T., 2004b, Football injuries during FIFA tournaments and the Olympic Games, 1998-2001. *American Journal of Sports Medicine*, **32**, S80-89.

Junge, A., Langevoort, G. and Pipe, A., 2006, Injuries in team sport tournaments during the 2004 Olympic Games. *American Journal of Sports Medicine*, **34**, pp. 565-76.

McMaster, W.C. and Walter, M., 1978, Injuries in soccer, *American Journal of Sports Medicine*, **6**, pp. 354-357.

Morgan, B. and Oberlander, A., 2001, An examination of injuries in major league soccer: The inaugural season. *American Journal of Sports Medicine*, **29**, pp. 426.

Ostojic, S.J., 2003, Comparing sports injuries in soccer: Influence of a positional role. *Sports Research in Sports Medicine*, **11**, pp. 203-208.

Price, R., Hawkins, R. and Hulse, M., 2004, The Football Association medical research programme: an audit of injuries in academy youth football. *British Journal of Sports Medicine*, **38**, pp. 466-71.

Rahnama, N., Reilly, T. and Lees, A., 2002, Injury risk associated with playing actions during competitive soccer. *British Journal of Sports Medicine*, **36**, pp. 354-359.

Wong, P. and Hong, Y., 2005, Soccer injury in the lower extremities, *British Journal of Sports Medicine*, **39**, pp. 473-482.

Yoon, Y.S., Chai, M. and Dong, W.S., 2004, Football injuries at Asian tournaments. *American Journal of Sports Medicine*, **32**, pp. 36S-42S.

Part IV

Body Composition and Nutrition

CHAPTER EIGHTEEN

Validation of a new anthropometric equation for the prediction of body composition in elite soccer players

L. Sutton, M. Scott and T. Reilly

Research Institute for Sport and Exercise Sciences, Liverpool John Moores University, Liverpool, UK

1. INTRODUCTION

Achieving an appropriate body composition is considered a key aspect of preparation for performance in professional soccer. Excess fat mass acts as a dead weight in activities in which the body is lifted against gravity. This added weight has an adverse effect upon general locomotion and soccer-specific skills such as jumping to head the ball or contest possession in the air (Reilly, 1996). A single assessment of body composition may be used alongside physiological tests as a measure of preparedness for competition. Furthermore, repeat assessments are useful in evaluating the effects of changes to exercise and/or dietary regimens on body composition status.

In the absence of a 'gold standard' laboratory method, traditional anthropometric measures are often used to assess body composition in soccer players. The most common field-based technique for the assessment of body composition, in particular the fat compartment, is the use of skinfold thicknesses determined by the caliper method, as reported in the majority of studies in which soccer players' body composition has been measured. In most cases, a regression equation is used to convert skinfold thickness values into an estimation of total body fat. Throughout the United Kingdom, the most commonly-used equation is that of Durnin and Womersley (1974), whilst the body mass index (BMI) tends to be used in population studies.

The Steering Groups of the British Olympic Association (Reilly *et al.*, 1996) advised against the conversion of skinfold measures into total percent body fat, given the inherent error associated with the use of generalised skinfold equations, especially in athletic populations. Instead, it was advised that the sum of skinfolds (ΣSKF) be used as a measure of subcutaneous adiposity, with changes in ΣSKF indicative of changes in body composition. This recommendation could be extended to the measurement of adiposity in soccer players in the absence of a sport-specific formula for estimating percent body fat. For use in situations where an estimation of percent body fat is preferred, Wallace *et al.* (2009) proposed a new soccer-specific skinfold equation:

Percent body fat = 5.011 + (0.175 x abdominal) + (0.182 x thigh) + (0.136 x calf)

using measurement techniques and landmarks described by the International Society for the Advancement of Kinanthropometry (Marfell-Jones *et al.*, 2006).

The main aim of the present study was to establish the validity of the new prediction equation and compare it to the traditional generalised skinfold equation of Durnin and Womersley (1974), using dual-energy x-ray absorptiometry (DXA) as the reference method. The second aim of the study was to assess the efficacy of the body mass index (BMI) and ΣSKF methods as indicators of adiposity.

2. MATERIALS AND METHODS

2.1 Participants

Seventeen senior soccer players of varying ethnicity were recruited from three clubs competing in the English Premier League. The age, mass and height of the participants were 24.3 ± 4.4 years, 1.81 ± 0.08 m and 82.1 ± 7.6 kg (mean ± SD), respectively.

2.2 Anthropometry

Height and mass (Seca 702, Seca GmbH & Co.KG, Hamburg, Germany) were recorded and BMI was calculated. Each player underwent a whole-body DXA scan (Hologic QDR Series Discovery A, Massachusetts, USA) and skinfold measurements at the biceps, triceps, subscapular, suprailiac (also termed iliac crest) and anterior thigh sites using Harpenden calipers (Baty International, West Sussex, UK). The equations of Durnin and Womersley (1974) and Wallace *et al.* (2009) were used to predict percent body fat. Percent body fat determined by DXA was used as the criterion value. The sum of the four sites (Σ4) recommended by Durnin and Womersley (1974) and the five (Σ5 = Σ4 + anterior thigh) recommended by Reilly *et al.* (1996) were calculated.

2.3 Statistical analysis

Body fat values obtained by the equations of Durnin and Womersley (1974) and Wallace *et al.* (2009) were assessed against DXA values using paired *t*-tests and the 95% limits of agreement method described by Bland and Altman (1986). The relationships between DXA-determined percent body fat and BMI, ΣSKF and the two skinfold equations were assessed using Pearson's correlation coefficient with 95% confidence intervals.

3. RESULTS

Data are presented as mean ± standard deviation. The criterion DXA-determined percent body fat was $10.7 \pm 1.9\%$. Estimations of percent fat by the traditional equation of Durnin and Womersley (1974) showed significantly high systematic bias compared to values derived from DXA. In contrast, the new equation of Wallace *et al.* (2009) slightly underestimated percent body fat. When the two prediction equations were compared, the equation of Wallace *et al.* (2009) demonstrated lower bias, a smaller standard deviation and narrower limits of agreement (Table 1).

Table 1. An assessment of bias and agreement of two skinfold equations for the prediction of percent body fat against DXA criterion values in professional soccer players.

Skinfold equation	Systematic bias $(\bar{x} \pm SD)$	t-test	95% limits of agreement (range)
Durnin and Womersley (1974)	+3.2 ± 1.6 %	$P < 0.001$	0.1-6.3 (6.2) %
Wallace *et al.* (2009)	-0.8 ± 1.0 %	$P < 0.001$	-2.7-1.1 (3.8) %

Of the two equations for the prediction of percent body fat, that of Wallace *et al.* (2009) showed the slightly stronger correlation with DXA-derived reference values (Table 2), although it should be noted that there is a certain amount of overlap between the respective confidence intervals. The ΣSKF method demonstrated strong positive correlations with DXA values, with the $\Sigma5$ showing a stronger relationship than the $\Sigma4$ skinfolds. No significant correlation was evident between BMI and percent body fat determined by DXA ($P = 0.166$).

Table 2. Correlations of field-based methods of body composition assessment with percent fat determined by DXA in professional soccer players.

Method of body composition assessment	r	95% CI
$\Sigma5 = 43.6 \pm 9.9$ mm	0.91	0.77 – 0.97
Wallace *et al.* (2009)	0.87	0.67 – 0.95
$\Sigma4 = 34.0 \pm 7.3$ mm	0.85	0.61 – 0.94
Durnin and Womersley (1974)	0.83	0.57 – 0.94
BMI = 25.1 ± 1.4	0.35	-0.15 – 0.71

CI = confidence interval.

4. DISCUSSION

Previous research in which DXA has been employed to model body composition has shown that elite, highly-trained soccer players differ from their non-athlete counterparts in both segmental and total body composition (Wittich *et al.*, 1998; Calbet *et al.*, 2001; Wittich *et al.*, 2001). Assuming an adequate diet, development of bone and muscle mass would be anticipated, given the physical demands and loading patterns on the soccer player during training and competition, and the accumulation of body fat should be limited (Shephard, 1999). These findings suggest that anthropometric equations generated using data from the 'general population' may not be appropriate for use with professional soccer players. Indeed, whilst the skinfold equation of Durnin and Womersley (1974) has been validated for use in the general population, its use in professional soccer groups cannot be recommended based on the present study. The results show an unacceptably high overestimation of percent body fat.

The greater accuracy, agreement with criterion values and lower variability shown by the soccer-specific equation of Wallace *et al.* (2009) provide support for the use of population-specific prediction equations. However, the significant bias was still cause for concern. This led the authors to combine the original group of Wallace *et al.* (2009) with players from the present sample to create a larger professional soccer group. The procedure of Wallace *et al.* (2009) was repeated, producing the following soccer-specific equation, which explained 78.4% variance in DXA-derived percent body fat:

$$Percent\ body\ fat = 5.174 + (0.124'\ thigh) + (0.147'\ abdominal) + (0.196'\ triceps) + (0.130'\ calf)$$

(Reilly *et al.*, 2009)

The larger sample size for generating the new equation provided a stronger statistical model and the preliminary work of Reilly *et al.* (2009) indicated good cross-validation predictive power. However, further validation work with a separate male soccer sample is warranted in order to assess whether the new soccer equation provides a consistently more suitable field-based method of assessing body fat in male soccer players than existing methods.

Older equations, such as that of Durnin and Womersley (1974), tend to be derived from reference body density and fat values obtained by the traditional hydrodensitometry method. The two-compartment (fat and fat-free) hydrodensitometry model of body composition is subject to a number of assumptions, many of which may be violated in athletic populations. It is assumed that the density of the fat-free mass is constant both between and within individuals. In addition to differing from non-soccer players, body composition is found to vary within the individual soccer player. Changes in bone mass due to physical activity are site-specific, meaning adaptations occur in the areas at which increased loading and mechanical strain occurs (Uzunca *et al.*, 2005). Due to the loading pattern and lower-body dominance of soccer, it would therefore be expected that the density of the lower-body mineral compartment increases in soccer players, as has been found previously in the lower limbs and lumbar spine (Calbet *et al.*, 2001). Similarly, the mineral-free compartment of the fat-free mass is not constant across the body, given that soccer players show a more pronounced muscular development in the lower body, in particular the hip flexors, quadriceps, hamstrings and ankle plantar- and dorsi-flexors (Reilly, 1996). Dual-energy x-ray absorptiometry and other indirect techniques derived from this method are not affected by this limitation.

The equations of Wallace *et al.* (2009) and Reilly *et al.* (2009) incorporate skinfold sites from the abdomen and lower body, whereas the sites included in the generalised equation of Durnin and Womersley (1974) are located on the upper body. In an assessment of various skinfold sites, Eston *et al.* (2005) found the anterior thigh to have the highest correlation with DXA-determined body fat and the combination of the thigh and calf to explain more variance in percent fat than the traditional Σ4 skinfolds. The authors concluded that lower-body skinfold sites are more highly related to percent fat in healthy individuals than are upper-body sites. This finding, in addition to the fact that the soccer-specific equations were produced using data collected from professional soccer players, using a three-compartment model of body composition, may explain its greater predictive ability.

The results of the present study showed the Σ5 skinfolds recommended by Reilly *et al.* (1996) to be an acceptable means of assessing adiposity in professional soccer players. The Σ5 showed a stronger relationship with DXA-determined body fat than the Σ4 skinfolds, providing further support for the inclusion of skinfold measurements from the lower body. The ΣSKF method was markedly superior to BMI as a measure of adiposity in the present sample. Body mass index is a 'weight-for-height' index and, therefore, does not take into account the composition of the body. To illustrate the significance of this limitation, if the limits set by the World Health Organisation (2000) were applied in the present sample, 11 of the 17 soccer players would be classified as 'overweight' or 'pre-obese'. In fact, the DXA-determined body fat for those 11 players ranged from 7.9 to 14.2%. These values fall within Wilmore and Costill (1999)'s recommended range of 6-14% body fat for soccer players. In accordance with the results of Wallace *et al.* (2009), BMI was found to have no correlation with DXA-determined body fat in the present sample and cannot be recommended for the assessment of adiposity in professional soccer players.

5. CONCLUSIONS

The lack of transferability of the generalised equation of Durnin and Womersley (1974) provides further support for the development of population-specific anthropometric equations, especially for use with highly-trained athletes. Specific regression equations should be developed using the target group and validated against 'gold standard' laboratory methods of body composition assessment such as DXA. Whilst showing promise, the equation of Wallace *et al.* (2009) has been improved in the new equation of Reilly *et al.* (2009) but requires further validation before it can be recommended unequivocally for the prediction of body fat in professional male soccer players. The recommendation of Reilly *et al.* (1996) to use the sum of skinfold sites, including the anterior thigh, without subsequent conversion to percent body fat, can be recommended for use as an indicator of adiposity in the field-based assessment of body composition in elite athletes.

References

Bland, J.M. and Altman, D.G., 1986, Statistical methods for assessing agreement between two methods of clinical measurements. *Lancet*, **i**, pp. 307-310.

Calbet, J.A., Dorado, C., Díaz-Herrera, P. and Rodríguez-Rodríguez, L.P., 2001, High femoral bone mineral content and density in male football (soccer) players. *Medicine and Science in Sports and Exercise*, **33**, pp. 1682-1687.

Durnin, J.V.G.A. and Womersley, J., 1974, Body fat assessed from total body density and its estimation from skinfold thickness: measurements of 481 men and women aged from 16-72 years. *British Journal of Nutrition*, **32**, pp. 77-97.

Eston, R.G., Rowlands, A.V., Charlesworth, S., Davies, A. and Hoppitt, T., 2005, Prediction of DXA-determined whole body fat from skinfolds: importance of including skinfolds from the thigh and calf in young, healthy men and women. *European Journal of Clinical Nutrition*, **59**, pp. 695-702.

Marfell-Jones, M., Olds, T., Stewart, A. and Carter, J.E.L., 2006, *International Standards for Anthropometric Assessment* (Potechefstroom: International Society for the Advancement of Kinanthropometry).

Reilly, T., 1996, Fitness assessment. In *Science and Soccer*, edited by Reilly, T. (London: E and FN Spon), pp. 25-47

Reilly, T., George, K., Marfell-Jones, M., Scott, M., Sutton, L. and Wallace, J., 2009, How well do skinfold equations predict percent body fat in elite soccer players? *International Journal of Sports Medicine*, in press.

Reilly, T. Maughan, R.J. and Hardy, L., 1996, Body fat consensus statement of the steering groups of the British Olympic Association. *Sports Exercise and Injury*, **2**, pp. 46-49.

Shephard, R.J., 1999. Biology and medicine of soccer: an update. *Journal of Sports Sciences*, **17**, pp. 757-786.

Uzunca, K., Birtane, M., Durmus-Altun, G. and Ustun, F., 2005, High bone mineral density in loaded skeletal regions of former professional football (soccer) players: what is the effect of time after active career? *British Journal of Sports Medicine*, **39**, pp. 154-157.

Wallace, J., Marfell-Jones, M., George, K. and Reilly, T., 2009, A comparison of skinfold thickness measurements and dual-energy x-ray absorptiometry analysis of percent body fat in soccer players. In *Science and Football VI*, edited by Reilly, T. and Korkusuz, F. (London: Routledge), pp. 364-369.

Wilmore, J.H. and Costill, D.L., 1999, *Physiology of Sports and Exercise*, 2nd ed. (Champaign: Human Kinetics).

Wittich, A., Mautalen, C.A., Oliveri, M.B., Bagur, A., Somoza, F. and Rotemberg, E., 1998, Professional football (soccer) players have a markedly greater skeletal mineral content, density and size than age- and BMI-matched controls. *Calcified Tissue International*, **63**, pp. 112-117.

Wittich, A., Oliveri, M.B., Rotemberg, E. and Mautalen, C., 2001, Body composition of professional football (soccer) players determined by dual x-ray absorptiometry. *Journal of Clinical Densitometry*, **4**, pp. 51-55.

World Health Organisation, 2000, Obesity: preventing and managing the global epidemic. Report of a WHO Consultation. *WHO Technical Report Series 894.* (Geneva: World Health Organisation.)

CHAPTER NINETEEN

Somatotype and body composition in Portuguese youth soccer players

B. Salgado, S. Vidal, S. Silva, R. Miranda, R. Deus, R. Garganta,
J. Maia, A. Rebelo and A. Seabra

Faculty of Sports, University of Porto, Porto, Portugal

1. INTRODUCTION

Somatotype is an overall description of physique, a continuum from leanness and linearity to roundness and heaviness. Somatotype is expressed by a three-number rating that represents the components of endomorphy (fatness), mesomorphy (musculoskeletal development) and ectomorphy (linearity). On the other hand, body composition aims at the division of body weight into several compartments, describing components such as body fat and lean body mass in a two-compartmental model.

Physique and body composition play a role in the selection of young athletes and are associated with the success of youth athletes' responses to training and competition. It has also been suggested (Wilmore and Costill, 1999) that physical characteristics are essential items in an efficient organization of a soccer team for reaching a successful resolution of matches.

Assessing the differences among youth soccer players who play in different positions is not a trivial task, depending on the match, the strategy, the opponent, and the demands on every player which sometimes change from match to match (Gil *et al.*, 2007). The aims of the present study were to determine somatotype components and body composition of youth Portuguese soccer players, and detect possible differences according to their playing position.

2. METHODS

A sample of 187 soccer players aged 17-18 years was evaluated. The players were classified according to their playing position: goalkeepers, full-backs, centre-backs, midfielders and forwards (Table 1).

The somatotype components were estimated using Heath and Carter's (1967) method based on 10 somatic measurements at various body sites. Body composition was assessed by using the equation of Boileau *et al.* (1985).

Differences by playing positions were tested with factorial analysis of variance. The Scheffé test for multiple post-hoc comparisons was used. All data analysis was done in SPSS 15.0.

3. RESULTS and DISCUSSION

3.1 Anthropometry

Body mass and height values are shown in Table 1. Mean values for junior soccer players were 70.5 kg and 1.75 m for mass and height respectively. Goalkeepers and centre-backs were the heaviest and tallest players.

These results are slightly lower than those reported in most recent studies with soccer players of the same age (Casáis *et al.*, 2004; Tschopp *et al.*, 2004; Valtueña *et al.*, 2006; Gil *et al.*, 2007) (See Table 2).

Table 1. Characteristics of junior soccer players according to their playing position (Mean ± *SD*).

	Playing positions					
	All	Goalkeepers	Full-backs	Centre-backs	Midfielders	Forwards
n	187	17	26	27	73	44
Body mass (kg)	70.5 ± 7.9	76.4 ± 9.9	69.8 ± 6.1	74.8 ± 7.4	68.8 ± 7.5	68.9 ± 7.3
Height (m)	1.75 ± 0.06	1.77 ± 0.05	1.74 ± 0.06	1.81 ± 0.05	1.75 ± 0.06	1.73 ± 0.06

Table 2. Recent studies of junior soccer players (Mean ± *SD*).

Author	*n*	Body mass (kg)	Height (m)
Casáis *et al.* (2004)	318	72.5 ± 7.4	1.76 ± 0.06
Gil *et al.* (2007)	126	74.0 ± 1.5	1.78 ± 0.02
Tschopp *et al.* (2004)	48	73.9 ± 6.5	1.79 ± 0.01
Valtueña *et al.* (2006)	46	74.2 ± 6.6	1.80 ± 0.06
Present study	187	70.5 ± 7.9	1.75 ± 0.06

3.2 Somatotype

Somatotype components by playing position are shown in Table 3. Mean somatotype of the youth soccer players was 3.03-4.78-2.55, thus falling within the category balanced-mesomorph.

Goalkeepers, centre-backs and midfielders were reported as balanced-mesomorph. Full-backs and forwards were matched to an endo-mesomorph profile. With the exception of the endomorph component, no significant differences were found between playing positions.

Goalkeepers presented the highest endomorphy values compared to midfielders (*P*=0.038) and forwards (*P*=0.031). In addition, the centre-backs tended to present high values of ectomorphy.

Table 3. Somatotype of junior soccer players according to their playing position (Mean ± *SD*).

Playing position	Endomorphy	Mesomorphy	Ectomorphy	Somatochart
	Somatotype			
1 Goalkeepers	3.79 ± 1.4[12]	5.23 ± 1.2	2.06 ± 0.9	
2 Full-backs	3.08 ± 0.9	4.89 ± 1.0	2.45 ± 0.8	
3 Centre-backs	3.04 ± 1.2	4.59 ± 0.9	2.84 ± 0.9	
4 Midfielders	2.93 ± 1.0	4.60 ± 0.9	2.66 ± 1.0	
5 Forwards	2.85 ± 0.8	4.94 ± 1.1	2.45 ± 0.9	
O All	3.02 ± 1.0	4.78 ± 1.0	2.55 ± 0.9	
F	3.040	1.966	2.271	
P	0.019	0.102	0.063	

[1] Goalkeepers vs. midfielders ($P < 0.05$)
[2] Goalkeepers vs. forwards ($P < 0.05$)

3.3 Body composition

The assessment of body composition showed similar results regarding the amount of body fat (Table 4). The average estimate of percent body fat among these youth soccer players was 16.1 (± 4.3) %. Thus, goalkeepers had significantly higher amounts of body fat than midfielders ($P=0.010$) and forwards ($P=0.004$).

Table 4. Body composition of junior soccer players according to their playing position (Mean ± *SD*).

	Goalkeepers	Full-backs	Centre-backs	Midfielders	Forwards	F	P
	Playing Positions						
Fat (%)	19.7 ± 5.0[12]	16.9 ± 4.6	16.2 ± 5.0	15.6 ± 3.9	15.0 ± 3.1	4.5	0.002
Fat (kg)	15.3 ± 5.6[12]	11.9 ± 3.7	12.8 ± 4.4	10.9 ± 3.4	10.4 ± 2.7	6.3	0.001
LBM (kg)	61.0 ± 6.4	57.9 ± 4.5	62.6 ± 5.8[3]	57.9 ± 5.8	58.4 ± 5.8	4.2	0.003

[1] Goalkeepers vs. midfielders ($P < 0.05$)
[2] Goalkeepers vs. forwards ($P < 0.05$)
[3] Centre-backs vs. midfielders ($P < 0.05$)

Our results showed higher percent fat mass than all the reports on junior soccer players in the recent literature (Casáis *et al.*, 2004; Tahara *et al.*, 2006; Valtueña *et al.*, 2006; Gil *et al.*, 2007) (See Table 5).

Table 5. Percent body fat of junior soccer players (Mean ± *SD*).

Author	n	Equations	Fat (%)
Tahara *et al.* (2006)	72	Brozek *et al.*	9.6 ± 3.0
Valtueña *et al.* (2006)	46	Faulkner	10.1 ± 0.8
Gil *et al.* (2007)	126	Faulkner	11.6 ± 0.2
Casáis *et al.* (2004)	318	Matiegka	11.7 ± 1.4
Present study	187	Boileau *et al.*	16.1 ± 4.3

4. CONCLUSIONS

On average, forwards were the shortest players, followed in order by full-backs, midfielders, goalkeepers and centre-backs. The body mass of the full-backs was, on average, midway between that of midfielders and forwards, on the one hand, and centre-backs and goalkeepers, on the other hand. Junior soccer players tended to be more mesomorphic and less endomorphic and ectomorphic.

These results support the following: (1) notwithstanding specific demands of training and competition according to playing position, no substantive differences were found among Portuguese junior soccer players. There may be selective pressure towards a relatively homogeneous physique and body composition values at this level of competition. (2) Although different predictive equations produce diverse body fat values, it is important to report our greater percent of body fat when compared to other studies. This observation may support differences in training demands and nutritional habits.

References

Boileau, R., Lohman, T. and Slaughter, M., 1985, Exercise and body composition of children and youth. *Scandinavian Journal of Sports Science*, **7**, pp. 17-27.
Carter, J. and Heath, B., 1990, Somatotyping - Development and Applications (Cambridge: Cambridge University Press).
Casáis, L., Salgado, J., Lago, E. and Peñas, C., 2004, Relacion entre parametros antropometricos y manifestaciones de fuerza y velocidad en futebolistas en edades de formation. III Congreso de la Asociación Española de Ciencias del Deporte. Valencia.
Gil, S., Ruiz, F., Irazusta, A., Gil, J. and Irazusta, J., 2007, Selection of young players in terms of anthropometric and physiological factors. *Journal of Sports Medicine and Physical Fitness*, **47**, pp. 25-32.
Heath, B. and Carter, J., 1967, A modified somatotype method. *American Journal of Physical Anthropology*, **27**, pp. 57-74.
Tahara, Y., Moji, K., Tsunawake, N., Fukuda, R., Nakayama, M., Nakagaichi, M.,

Komine, T., Kusano, Y. and Aoyagi, K., 2006, Physique, body composition and maximum oxygen consumption of selected soccer players of Kunimi High School, Japan. *Journal of Physiological Anthropology*, **25**, pp. 291-297.

Tschopp, M., Held, T. and Marti, B., 2004, Four-year development of physiological factors of junior elite soccer players aged 15-23 years. (Part III: physiological and kinanthropometry.) *Journal of Sports Sciences*, **22**, pp. 564-565.

Valtueña, J., González-Gross, M. and Sola, R., 2006, Status en hierro de jugadores de fútbol y baloncesto de la categoría junior. *Revista Internacional de Ciencias del Deporte*, **22**, pp. 57-68.

Wilmore, J. and Costill, D., 1999, *Physiology of Sports and Exercise*, 2nd ed. (Champaign: Human Kinetics), pp. 490-507.

CHAPTER TWENTY

Effects of sodium bicarbonate supplementation on performance related to association football

P.A. Ford[1], S. Cousins[1] and J. Johnstone[2]

[1]School of Health and Bioscience, University of East London, UK
[2] Sport, Health and Exercise Sciences, University of Hertfordshire, UK

1. INTRODUCTION

Outfield players travel a total distance of ~11 km during a game of association football (soccer) in the form of walking, jogging, and sprinting (Bangsbo and Lindquist, 1992; Bangsbo 1994; Reilly *et al.*, 2000). Although aerobic processes dominate adenosine triphosphate (ATP) resynthesis, during the game players must utilise energy sources from anaerobic metabolism as well to meet the intermittent demands of activity. Players perform a sprint every ~90 s, complete high intensity work every ~30 s, and adapt their movements every ~6 s (Reilly, 1997). An outcome from this process of repetitive high-intensity activities is that lactate production increases which results in acidosis in skeletal muscle and blood (Roberg, 2001).

Bouissou *et al.* (1989) reported that an intra-muscular decrease in pH (acidosis) caused the slowing down of action potential and mean power frequency. Moreover, this decrease in intra-muscular pH placed a limit on the rate of ATP resynthesis, and inhibition of calcium ion release from the sarcoplasmic reticulum, inhibiting muscle contraction (McNaughton and Thompson, 2001). Although, Meyer *et al.* (1983) and Stephens *et al.* (2002) stated that intra-cellular pH is not the sole factor in fatigue, Bangsbo (1994) highlighted the detrimental effect this may have on performance in soccer.

With the evident degrading effects on performance, there has been much research into the way in which the onset of fatigue can be delayed in numerous sports. It seems that this objective can be achieved by increasing the blood buffering capacity and rate of hydrogen ion (H^+) efflux, which can be augmented simply by oral ingestion of sodium bicarbonate ($NaHCO_3$), and subsequently contribute to enhanced/maintained performance (McNaughton *et al.*, 1999; Stephens *et al.*, 2002; Price *et al.*, 2003). However, to the authors' best knowledge the effects of this supplement have not been investigated for soccer. This may in part be due to there being several studies that have shown no effect on performance outcomes using $NaHCO_3$ ingestion (Potteiger *et al.*, 1996; Aschenbach *et al.*, 2000; Marx *et al.*, 2002). Similarly, it may also be attributed to the fact that due to the prolonged nature of the game any augmentation is worthless because of dynamic homeostatic corrections (Brien and McKenzie, 1989), as well as many participants suffering from short-term stomach complaints after ingestion (Avedisian, 1996). An additional complexity for this subject is the difficulty in conducting an accurate and reliable exercise protocol that replicates the demands of a game situation.

The aim of this study is to observe whether the use of $NaHCO_3$ supplementation is effective in physical performance related to soccer.

2. METHODS

2.1 Participants

Twelve healthy, male, high-level amateur football participants (age 20.3 ± 1.50 years; stature 1.78 ± 0.04 m; body mass 76.83 ± 7.53 kg; $\dot{V}O_2$ max 54.63 ± 3.03 $ml.kg^{-1}.min^{-1}$; mean $\pm$ SD) were recruited for this study. These participants were recruited on the basis that they had already performed and knew the basic movements that the Loughborough Intermittent Shuttle Test (LIST) protocol encompasses (Nicholas *et al.*, 2000). Participants gave informed consent to participate in the study, which had received University of Brighton Ethics Committee approval.

2.2 Equipment and procedures

Descriptive data were collected from the participants using a stadiometer and scales (Seca Ltd, Birmingham, England). Likewise, a multi-stage fitness test was performed to predict maximal aerobic capacity (Ramsbottom *et al.*, 1988). The study was a repeated measures design, where the participants completed the LIST protocol twice. There was a 7-day gap between the tests. Prior to each test, participants received either the $NaHCO_3$ supplement (0.3 $g.kg^{-1}$ of body mass) or the control placebo (2 g of sodium chloride); following blind random selection using random number generation function on SPSS Version 13.0 (SPSS Inc, Chicago, IL). The supplements were consumed within gelatine capsules (Shearman *et al.*, 2003), with participants ingesting a maximum of 4 capsules every 15 min, 90 min prior to testing. The participants reported to the designated test site 30 min before testing, and were briefed on both occasions about the format of the protocol. Participants had abstained from physical exertion 48 hours prior to testing, were assessed at the same time of day, and had maintained similar food and drink consumption 24 hours prior to both tests. Upon arrival, participants were fitted with Polar Series 610i Heart Rate monitors (Polar Pacer, Port Washington, NY, USA), and had resting values recorded, along with values for rating of perceived exertion (RPE) and gastrointestinal discomfort (GID). Perceived exertion (RPE) was assessed using the Borg Scale (Borg, 1962), whereas GID was assessed using a novel Likert scale (1 = Nothing; 8 = Severe discomfort). These variables were also recorded after the protocol was completed. Prior to the start of the protocol, participants completed a 15-min standardised warm-up, which consisted of jogging, stretching, and striding.

The LIST protocol involved completing 5 interval exercise periods lasting 15 min each, followed by a run to exhaustion that normally lasted between 3-15 min. Each period involved several lengths between two cones 20 m apart: 3 lengths at walking pace, 1 length sprint, 4 s static recovery period, 3 lengths jogging (~55% $\dot{V}O_2$ max), and 3 lengths cruising (~95% $\dot{V}O_2$ max). The number of sets the participants were required to complete during the 15-min exercise period was related to their own predicted $\dot{V}O_2$ max, but was between 10-12 (Nicholas *et al.*, 2000). After each bout of exercise the participants then took a 3-min rest. Participants then repeated these work patterns for a further four bouts and corresponding rest periods to complete Part A of this protocol. The participants then performed Part B of the protocol, the run to exhaustion, which involved one length at jogging speed (55% $\dot{V}O_2$ max), and the return length at cruising speed (95% $\dot{V}O_2$max). Participants continued this until they could not maintain their required pace. Each participant was recorded by an individual investigator throughout the protocol to quantify workload. The protocol and measurement technique used have been shown previously to be accurate and reliable (Nicholas *et al.*, 2000).

2.3 Statistical analysis

All the descriptive results for this study are presented as mean ± one standard deviation. Data were analysed using SPSS Version 13.0. Homogeneity of variance was assessed by means of the Kolmogorov-Smirnov Test, with Lilliefors Significance Correction. Data-sets were normally distributed, and therefore, differences between variables within the protocols following either the $NaHCO_3$ or the control supplementation were measured using a paired t-test. Statistical significance was accepted as $P<0.05$.

3. RESULTS

Participants' mean distance covered during the controlled Part A of the repeated protocols following the $NaHCO_3$ and control supplementation was 10.73 ± 0.33 km and 10.78 ± 0.29 km, respectively, which was not significantly different ($P<0.05$). Though there were individual fluctuations in responses (see Figure 1), paired t-tests showed that the distance covered during the run to exhaustion (Part B) was significantly greater after the $NaHCO_3$ supplement compared to the control supplement ($P<0.05$). This difference during the LIST protocol represented a significant increase in the total distance covered by subjects following $NaHCO_3$ compared to the control supplementation ($P<0.05$). See Table 1 for data. There were no significant differences in the participants' mean heart rate, RPE, and gastrointestinal discomfort during any part of the repeated measures study ($P<0.05$). See Table 2.

Table 1. LIST protocol variables following NaHC0$_3$ and control supplementation.

LIST Protocol	NaHC0$_3$			Control		
	Mean	SD	95% CI	Mean	SD	95% CI
Part A – Distance covered (km)	10.78	0.29	10.62-10.94	10.73	0.33	10.55-10.92
Part B – Distance covered (km)	0.83	0.3	0.68-0.98	0.75	0.3	0.58-0.93
Part A and B – Distance covered (km)	11.61	0.6	11.19-11.43	11.57	0.5	10.95-11.27

95% CI = 95% Confidence Interval

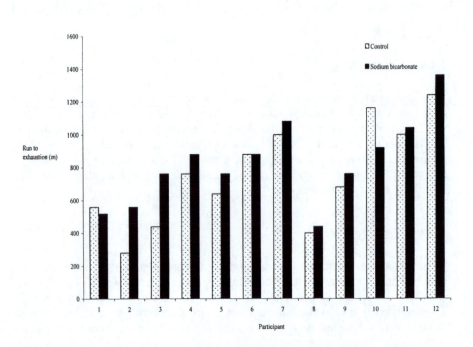

Figure 1. Individual differences in run to exhaustion following sodium bicarbonate and control supplementation.

Table 2. Heart rate (HR), rating of perceived exertion (RPE) and gastrointestinal discomfort (GID) at the end of respective LIST protocol stages following NaHCO3 and control supplementation.

LIST Protocol		*NaHCO3*			*Control*		
		Mean	SD	95% CI	Mean	SD	95% CI
Part A:	HR (beats.min^{-1})	172	7.02	168-176	172	13.8	164-180
	RPE	13	1.1	12.8-14.0	14	0.7	13.4-14.2
	GID	2.5	1.0	1.9-3.1	1.8	1.0	1.2-2.4
Part B:	HR (beats.min^{-1})	185	8.2	180-189	187	7.3	183-191
	RPE	18	0.8	17.1-18.1	18	0.7	17.4-18.2
	GID	2.5	1.0	1.9-3.1	2.0	0.7	1.6-2.4

95% CI = 95% Confidence Interval

4. DISCUSSION

To the authors' best knowledge this is the first study employing a prolonged intermittent protocol, similar to that of the demands of playing soccer, which showed performance was significantly enhanced following NaHCO3 supplementation.

Numerous authors have highlighted the detrimental effects acute and chronic fatigue can have on performance variables related to soccer (Sprague and Mann, 1983; Nummela *et al.*, 1994; Tupa *et al.*, 1995; Derrick *et al.*, 2002). Although the body has several compensatory strategies to help reduce the onset of fatigue, by artificially increasing buffering capacity and rate of intra-cellular H$^+$ efflux this point can be delayed (Gaitanos *et al.*, 1991). This is because the decrease in extra-cellular acidosis facilitates the efflux of H$^+$ out of the muscles into the blood using the diffusion gradient. This movement allows the pH of the sarcoplasm to be maintained, in that it becomes less acidic, and therefore means that it does not limit functioning in terms of metabolism and contraction for both short-term and prolonged exercise (Sutton *et al.*, 1981; Robertson *et al.*, 1986). Use of NaHCO3 has been observed to augment the body's bicarbonate reserves in terms of both acute and chronic methods for enhancing brief and prolonged high-intensity exercise (McNaughton and Thompson, 2001; Stephens *et al.*, 2002). The key concept of buffering involves the 'pK' of an acid, which is where dynamic equilibrium between protons leaving and re-attaching to acid molecules occurs. In summary, because NaHCO3 has a pK value of 7.4 after a series of chemical reactions that occur inside the body it is an effective blood buffer (Robergs, 2002).

Nonetheless, it seems that NaHCO3 supplementation has not been studied previously in soccer. Many researchers have struggled to measure the actual demands of soccer due to the fact it is unorthodox in movement requirements and involves non-cyclical movement patterns (Reilly, 1997). As mentioned previously, several investigations have shown positive effects on exercise performance in delaying the onset of fatigue through $NaHCO_3$ supplementation (McNaughton *et al.*, 1999; Stephens *et al.*, 2002). Specifically, Price *et al.* (2003) reported that $NaHCO_3$ supplementation can enhance performance for prolonged intermittent exercise, which is the nature of soccer play (Reilly, 1997). The supplement did not stop blood or muscle pH from dropping, but helped to maintain a less acidic internal environment for prolonged intermittent exercise to continue. Such results are the same as those in the present study, but the novelty of these current findings lies in that the protocol directly corresponds to the match demands of soccer.

There have also been negative findings using sodium bicarbonate supplementation (Potteiger *et al.*, 1996; Marx *et al.*, 2002). Aschenbach *et al.* (2000) found that performance in prolonged activity may not be enhanced, as the supplement only brings short-term alterations when homeostasis is in a state of flux. This was not apparent in the present study, as the exercise was over a prolonged period of time and performance was still enhanced. Similarly, other studies have also suggested that the ergogenic benefits of $NaHCO_3$ are hidden due to stomach complaints following its ingestion (Avedisian, 1996). This result was also not seen in the present study, as there were no significant differences in any of the subjects' perceived rating of gastrointestinal discomfort at any stage of the protocols.

In terms of limitations, measures of blood bicarbonate and pH were not taken due to budget and practical restraints, which meant that there is no direct mechanistic measure. Such a measure would have confirmed that the supplement caused the enhanced performance, rather than being a biological variance. Nevertheless, this was a similar experimental design to the previous controlled tests on $NaHCO_3$ that produced significant findings (McNaughton *et al.*, 1999; Price *et al.*, 2003). Similarly, another concern was that the LIST protocol might not fully replicate the physiological demands of playing in a game, and is based upon predicted maximal aerobic capacity (Ransbottom *et al.*, 1988). Nevertheless the participants subjectively reported that the test was related to the demands of a game, and the methods used have previously been reported as valid and reliable (Nicholas *et al.*, 2000). In summary, after acknowledging the accuracy and reliability of the measures the likelihood of a type I error occurring during this study is minimal. Perhaps the subject number for the study may have been too small, meaning a type II error may have occurred. The sample size for the present study was based upon a statistical power of 80% and an estimated effect size of 0.84 (McNaughton *et al.*, 1999). Yet, the results of this study actually indicate that there is a lower effect size of 0.25 for soccer. Furthermore, there is a plausible physiological rationale to explain why the changes occur between the $NaHCO_3$ and control supplementation within this repeated measures design.

The findings of the present study address enhancement of individual performance, through superior physical performance, but this does not necessarily mean that team performance will be enhanced as well. Perhaps as well as conducting a future study with a larger sample size, the implication of $NaHCO_3$

supplementation upon team performance could be assessed. Even though this supplement may be used by coaches and trainers to help avoid errors and maximise the chances of the team performing well, it may not enhance tactical play or high levels of skill performances. It simply means that the players are more prepared for meeting the demands of the game (Reilly and Williams, 2003).

5. CONCLUSION

The present findings suggest that $NaHCO_3$ supplementation may help to delay the onset of fatigue and enhance performance related to soccer. The implication of this study is that the supplement can be prescribed as an ergogenic tool for coaches and players during competition or training for prolonged intermittent activity.

References

Aschenbach, W., Ocel, J., Craft L., Ward, C., Sprangenburg, E. and Williams, J., 2000, Effect of oral sodium loading on high-intensity arm ergometry in college wrestlers. *Medicine and Science in Sports and Exercise,* **32**, pp. 669-675.

Avedisian, L., 1996, The effects of selected buffering agents on performance in the competitive 1600 meter run. *Microform Publications, Int'l Inst for Sport and Human Performance, University of Oregon Eugene, Ore., 1 microfiche (92 fr.): negative, ill.; 11x 15cm.*

Bangsbo, J., 1994, *Fitness Training in Football: A Scientific Approach* (Bagsværd: HO+Storm).

Bangsbo, J. and Lindquist, F., 1992, Comparison of various exercise tests with endurance performance during soccer in professional players. *International Journal of Sports Medicine,* **13**, pp. 125-132.

Borg, G.A.V., 1962, *Physical Performance and Perceived Exertion* (Lund: Gleerup), pp. 1-35.

Bouissou, P., Estrade, P.Y., Goubel, F., Guezennec, C.Y. and Serrurier, B., 1989, Surface EMG power spectrum and intramuscular pH in human vastus lateralis muscle during dynamic exercise. *Journal of Applied Physiology,* **67**, pp.1245-1249.

Brien, D.M. and McKenzie, D.C., 1989, The effect of induced alkalosis and acidosis on plasma lactate and work output in elite oarsmen. *European Journal of Applied Physiology and Occupational Physiology,* **58**, pp.797-802.

Derrick, T.R., Dereu, D. and Mclean, S.P., 2002, Impacts and kinematic adjustments during an exhaustive run. *Medicine and Science in Sports and Exercise,* **34**, pp. 998-1002.

Gaitanos, G.C., Nevill, M.E., Brooks, S. and Williams, C., 1991, Repeated bouts of sprint running after induced alkalosis. *Journal of Sports Sciences,* **9**, pp. 355-370.

Marx, J.O., Gordon, S.E., Vos, N.H., Nindl, B.C., Gomez, A.L., Volek, J.S., Pedro,

J., Ratamess, N., Newton, R.U., French, D.N., Rubin, M.R., Hakkinen, K. and Kraemer, W.J., 2002, Effect of alkalosis on plasma epinephrine response to high intensity cycle exercise in humans. *European Journal of Applied Physiology*, **87**, pp. 72-77.

McNaughton, L. and Thompson, D., 2001, Acute versus chronic sodium bicarbonate ingestion and anaerobic work and power output. *Journal of Sports Medicine and Physical Fitness*, **41**, pp. 456-462.

McNaughton, L., Dalton, B., and Palmer, G., 1999, Sodium bicarbonate can be used as an ergogenic aid in high-intensity, competitive cycle ergometry of 1 h duration. *European Journal of Applied Physiology and Occupational Physiology*, **80**, pp. 64-69.

Meyer, R.A., Kushmerick, M.J., Dillion, P.F. and Brown, T.R., 1983, Different effects of decreased intracellular pH on contractions in fast versus slow twitch muscle. *Medicine and Science in Sports and Exercise*, **B10**, pp. 116.

Nicholas, C.W., Nuttal, F.E. and Williams, C., 2000, The Loughborough Intermittent Shuttle Test: A field test that simulates the activity pattern of soccer. *Journal of Sports Sciences*, **18**, pp. 97-104.

Nummela, A., Rusko, H. and Mero, A., 1994, EMG activities and ground reaction forces during fatigued and no fatigued sprinting. *Medicine and Science in Sports and Exercise*, **26**, pp. 605-609.

Potteiger, J.A., Webster, M.J., Nickel, G.L., Haub, M.D. and Palmer, R.J., 1996, The effects of buffer ingestion on metabolic factors related to distance running performances. *European Journal of Applied Physiology and Occupational Physiology*, **72**, pp. 365-371.

Price, M., Moss, P. and Rance, S., 2003, Effects of sodium bicarbonate ingestion on prolonged intermittent exercise. *Medicine and Science in Sports and Exercise*, **35**, pp. 1303-1308.

Ramsbottom, R., Brewer, J. and Williams, C. 1988, A progressive shuttle run test to estimate maximal oxygen uptake. *British Journal of Sports Medicine,* **22**, pp. 141-144.

Reilly, T., 1997, Energetics of high-intensity exercise (soccer) with particular reference to fatigue. *Journal of Sports Sciences*, **15**, pp. 257-263.

Reilly, T., Bangsbo, J. and Franks, A., 2000, Anthropometric and physiological predispositions for elite soccer. *Journal of Sports Sciences*, **18**, pp. 669-683.

Reilly, T. and Williams, A.M., 2003, *Science and Soccer 2nd Ed.* (London: Routledge).

Robergs, R.A., 2001, Exercise-induced metabolic acidosis: Where do the protons come from? Sportscience, 5, *www.sportsci.org/jour/0102/rar.htm.*

Robergs, R.A., 2002, Explanation of bicarbonate and citrate buffering. *Journal of Exercise Physiology*, **5**, pp. 1-5.

Robertson, R.J., Falkel, J.E., Drash, A.L., Swank, A.M., Metz, K.F., Spungen, S.A. and LeBoeuf, J.R., 1986, Effect of blood pH on peripheral and central signals of perceived exertion. *Medicine and Science in Sports and Exercise*, **18**, pp. 114-122.

Shearman, J.P., Van Montfoort, M.C.E., Van Dieren, L. and Hopkins, W.G., 2003, Effects of ingestion of sodium bicarbonate, citrate, lactate, and chloride on sprint performance. *Medicine and Science in Sports and Exercise*, **35**, S269.

Sprague, P., and Mann, R.V., 1983, The effects of muscular fatigue on the kinetics of sprint running. *Track and Field Research Quarterly*, **54**, pp. 60-66.

Stephens, T.J., McKenna, M.J., Canny, B.J., Snow, R.J. and McConnell, G.K., 2002, Effect of sodium bicarbonate on muscle metabolism during intense endurance cycling. *Medicine and Science in Sports and Exercise*, **34**, pp. 614-621.

Sutton, J.R., Jones, N.L. and Toews, C.J., 1981, Effect of pH on muscle glycolysis during exercise. *Clinical Science*, **61**, pp. 331-338.

Tupa, V., Gusenov, F. and Mironenko, I., 1995, Fatigue influenced changes to sprinting technique. *Modern Athlete and Coach,* **33**, pp. 7-10.

CHAPTER TWENTY-ONE

Impact of combined carbohydrate ingestion on the metabolic responses to soccer-specific exercise in the heat

N. D. Clarke, B. Drust, D.P.M. MacLaren and T. Reilly

Research Institute for Sport and Exercise Sciences, Liverpool John Moores University, Liverpool, UK

1. INTRODUCTION

Soccer matches at major tournaments are regularly played in temperatures exceeding 30°C (FIFA World Cup 2002 and 2006 and UEFA Euro 2004), conditions that entail heat strain. Large amounts of water may be lost as a result of sweating during prolonged exercise in the heat and dehydration during exercise raises core temperature and increases cardiovascular strain (Sawka et al., 1985). The ingestion of fluid containing carbohydrate has been shown to offset dehydration, minimize disturbances in cardiovascular function, improve thermoregulation (Coyle and Coggan, 1984), maintain blood glucose concentration and improve performance (Davis et al., 1988) in moderate environmental temperatures.

Performing high-intensity intermittent exercise in the heat has been found to increase muscle glycogen utilization (Morris et al., 2005), fluid loss and cardiovascular stress compared to continuous exercise (Morris et al., 1998; Morris et al., 2005). The ingestion of a glucose solution during this type of exercise has been reported to be ineffective in attenuating the effects of dehydration and delaying the onset of fatigue (Morris et al., 2003). A possible reason for this occurrence is that previous studies (Davis et al., 1990; Jentjens et al., 2006) have indicated that the biological availability of fluid ingested during exercise in the heat is lower with a glucose drink compared with a combined glucose and fructose drink or water. Shi et al. (1995) demonstrated that the ingestion of a glucose and fructose solution resulted in greater water absorption than a solution of glucose only. These findings suggest that a glucose drink may be less effective for fluid replacement during exercise in the heat compared to a mixed carbohydrate drink. Therefore, ingesting a multi-carbohydrate drink may increase the intestinal absorption of fluid, thus minimising the impact of dehydration.

A potential limiting factor for the oxidation of exogenous carbohydrate is the rate of intestinal absorption of carbohydrate. It is thought that the intestinal glucose transporters (SGLT1) are saturated when the rate of glucose ingestion exceeds 1 $g \cdot min^{-1}$, which may explain why there is not a linear relationship between glucose ingestion rates and oxidation rates (Jeukendrup and Jentjens, 2000). However, Shi *et al.* (1995) suggested that the inclusion of two or three carbohydrates (glucose, fructose and sucrose) increases water and carbohydrate absorption despite increased osmolality. This effect was attributed to the separate transport mechanisms across the intestinal wall for glucose and sucrose. Therefore, when a solution containing a mixture of glucose and fructose is ingested, there is less competition for absorption compared with an isoenergetic amount of glucose. As a consequence there is the possibility of an increase in the amount of carbohydrate entering the bloodstream and its subsequent availability for oxidation, which in turn would spare muscle glycogen and delay the onset of fatigue, and improve performance. However, the majority of investigations into the effect of multi-carbohydrate solutions on carbohydrate oxidation have employed prolonged low-intensity exercise protocols (Jentjens *et al.*, 2004; Jentjens *et al.*, 2006) and the impact on exercise capacity or performance has not been studied.

The aim of this study was to investigate the effect of ingesting a multi-carbohydrate sports drink compared with a "glucose only" solution on metabolism and exercise capacity during soccer-specific exercise performed in the heat.

2. METHODS

Eleven male university soccer players of age: 27 ± 2 years; height: 1.78 ± 0.1 m; body mass: 76.1 ± 2 kg; $\dot{V}O_{2max}$: 63.1 ± 2 ml.kg^{-1}.min^{-1} participated in this study. All subjects provided written informed consent to participate, in accordance with Liverpool John Moores University's ethical procedures.

Subjects completed the full soccer-specific protocol, which was a modified version of that designed by Drust *et al.* (2000), on a motorised treadmill (H/P/Cosmos Pulsar 4.0, H/P/Cosmos Sports & Medical GmbH, Germany) on two occasions in an environmental chamber ($30.2\pm0.5°C$ and $45\pm4\%$ relative humidity). The soccer-specific protocol consisted of 90 min activity divided into 2 x 45 min identical periods, separated by a period of 15 min, representing half-time. Each 45-min period consisted of three 15-min blocks (Figure 1). The protocol consisted of the various exercise intensities that are included during competitive soccer matches (i.e. walking, jogging, cruising and sprinting). The proportions of these activities were based on the observations of Reilly and Thomas (1976). These activities were divided between walking and jogging. The proportion of time for each activity and corresponding speed was as follows: static pauses 3.8% (0 km.h^{-1}); walking 27.9% (4 km.h^{-1}); jogging 38.9% (12 km.h^{-1}); cruising 19.9% (15 km.h^{-1}); sprinting 9.5% (19 km.h^{-1}). The duration of each activity was determined by matching the proportions observed by Reilly and Thomas (1976) to the total time of the block, after the deduction of the total time for the treadmill speed changes had been made. The duration of each discrete bout was as follows: static pauses 8.0 s; walking 27.8 s; jogging 38.7 s; cruising 34.8 s; sprinting 9.4 s.

During one session 228 ± 6 ml of carbohydrate-electrolyte solution Still Lucozade Sport (6.6 g.100 ml^{-1} CHO, 49 mg.100 ml^{-1} Na, 296 ± 0.5 mOsm.kg^{-1}), GlaxoSmithKline, Gloucestershire, UK was consumed at 0, 15, 30, 45, 60 and 75 minutes of exercise (GLU). On another occasion 228 ± 7 ml of a multi-carbohydrate (fructose, dextrose, maltodextrin) solution (6.6 g.100 ml^{-1} CHO, 50 mg.100 ml^{-1} Na, 313 ± 0.6 mOsm.kg^{-1}, GlaxoSmithKline, Gloucestershire, UK) was consumed at the same time points (MIX). As a consequence, carbohydrate was ingested at a rate of 60 ± 0.5 g.h^{-1}. The trials were performed in a double-blind, counter-balanced fashion.

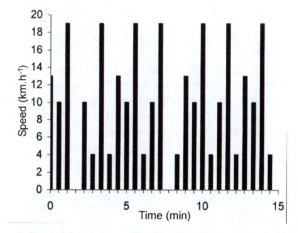

Figure 1. The 15-min activity profile of one block of the soccer-specific protocol..

To assess any changes in substrate oxidation rates, during a 2-min (10-12 min) period of each 15-min block, oxygen consumption ($\dot{V}O_2$) and carbon dioxide production ($\dot{V}CO_2$) were recorded using an on-line automated gas analyser (Metalyzer3B, Cortex Biophysic GmbH, Leipzig, Germany). These data were then used to calculate the rate of carbohydrate oxidation (Frayn, 1983).

After completing the soccer-specific protocol, subjects performed an anaerobic capacity test (Cunningham and Faulkner, 1969), which required the subject to run at a gradient of 20% and a speed of 12.8 km·h^{-1} until volitional exhaustion. The time began when the subject started running unsupported and stopped when he grabbed the handrails at the point of fatigue. This test was a measure of fatigue resistance to high-intensity exercise and has been shown to be both valid and reliable as a measurement tool (Thomas *et al.*, 2002).

Venous blood samples were taken from an antecubital vein in the forearm by a trained phlebotomist using the Vacutainer™ collection system (Becton Dickinson Vacutainer Systems Europe, Meylan, France). A blood sample was taken 30 min before exercise commenced, at half-time and at the completion of the protocol. All tubes were centrifuged and the plasma was frozen at -80°C for subsequent analysis. Plasma samples were analysed for glucose (Glucose oxidase, Instrumentation Laboratory, Monza, Italy).

A percutaneous needle biopsy sample of *vastus lateralis* was obtained approximately one week before the first trial. An additional biopsy was taken on completion of the soccer-specific protocol during each trial. After local anaesthesia 2 ml 0.5% Bupivacaine Hydrochloride (Marcain Polyamp, AstraZeneca, UK) and incision of the skin and muscle fascia, percutaneous muscle samples (~30 mg) were taken from the lateral vastus of the quadriceps femoris muscle using an automated procedure (Pro-Mag 2.2 Automatic Biopsy System, Manan Medical Products, USA) with a 14-gauge needle (ACN Biopsy needles, InterV, Denmark) in the distal to proximal direction. The biopsy was immediately frozen in liquid nitrogen and stored at -80°C for subsequent glycogen analysis. To determine the concentration of muscle glycogen the tissue was acid hydrolyzed allowing the glucose residues to be measured enzymatically (Powerwave X340, BioTek Instruments Inc, USA) as described by Lowry and Passonneau (1972) and expressed as "wet weight".

All variables were analysed using two-way ANOVAs with repeated measures with the exception of the time to exhaustion during the Cunningham and Faulkner test, sweat loss and changes in plasma volume, which were analysed using a one-way ANOVA. Results are reported as the mean ± the standard error of the mean (SEM) and a level of $P<0.05$ was considered statistically significant.

3. RESULTS

Muscle glycogen concentration (Figure 2) was significantly lower following the soccer-specific protocol compared with pre-exercise ($F_{1,7}=10.14$; $P<0.05$). However, the difference in utilization was not significant between the drinks ($F_{1,1}=1.42$; $P>0.05$): GLU (58.61 ± 2.7 mmol·kg wet weight^{-1}) and MIX (50.60 ± 6.9 mmol·kg^{-1} wet weight).

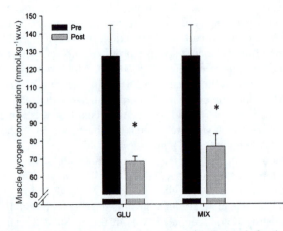

Figure 2. Mean±SEM muscle glycogen concentration pre-exercise (Pre) and after the soccer-specific protocol (Post). * significantly lower than pre-exercise.

The concentration of plasma glucose (Figure 3) was not significantly different between GLU and MIX ($F_{1,10}=1.13$; $P>0.05$). Also, plasma glucose concentration was not significantly different at the completion of the soccer-specific protocol (90 min) compared with at half-time (45 min) ($F_{1,10}=0.62$; $P>0.05$). Total carbohydrate oxidation (Figure 4) was not significantly affected by the different drink compositions ($F_{1,10}=0.05$; $P>0.05$). In addition, carbohydrate oxidation was similar during both halves of the protocol ($F_{1,10}=0.40$; $P>0.05$).

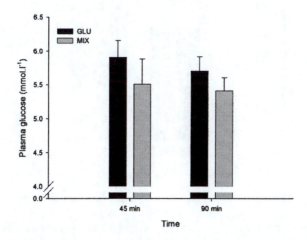

Figure 3. Mean±SEM plasma glucose concentration during the soccer-specific protocol.

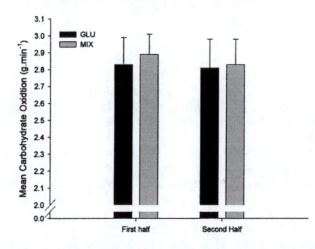

Figure 4. Mean±SEM carbohydrate oxidation during the soccer-specific protocol.

The mean duration of performance in the Cunningham and Faulkner test was: GLU: 77.11 ± 7.17 s; MIX: 83.04 ± 9.65 s (Figure 5). When exercise capacity was assessed, there was no significant difference between the two drinks ($F_{1,10}=1.98$; $P>0.05$).

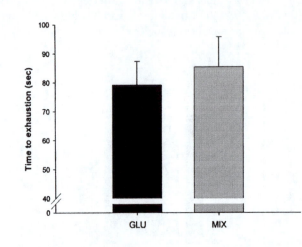

Figure 5. Mean±SEM time to exhaustion during Cunningham and Faulkner's test.

There were no significant differences in sweat loss between the two drinks, GLU (2.11 ± 0.2 kg) and MIX (2.10 ± 0.1 kg) ($F_{1,9}$=0.019; P>0.05). In addition, plasma volume changes (GLU: -1.91 ± 0.2%; MIX: -1.84 ± 0.3%) ($F_{1,8}$=1.447; P>0.05) were not significantly different between the trials with different types of drink.

4. DISCUSSION

The major findings of this study were that the ingestion of a solution containing glucose and fructose did not significantly alter exercise capacity during a high-intensity exercise test or the rate of total carbohydrate oxidation compared with the ingestion of a solution containing only glucose.

In the present study the lack of difference in exercise capacity between GLU and MIX may have been as a consequence of similar muscle glycogen content. The concentration of muscle glycogen has been shown to influence exercise capacity (Coyle and Coggan, 1984), and at the end of the soccer-specific protocol was similar irrespective of the fluid composition. Jentjens *et al.* (2006) recently demonstrated that in non-acclimated athletes exercising in the heat the ingestion of a solution containing glucose and fructose resulted in approximately 36% greater exogenous carbohydrate oxidation rates compared with the ingestion of a solution containing only glucose. In contrast, Wallis *et al.* (2005) reported that whilst the ingestion of carbohydrate significantly suppressed endogenous carbohydrate oxidation (muscle glycogen), there was no significant difference in the sparing of muscle glycogen between a maltodextrin drink and an isoenergetic maltodextrin plus fructose drink.

Plasma glucose concentration was also found to be similar for both trials, although slightly higher values were observed when the glucose only solution was ingested. This observation could be attributed to the fact that fructose must be converted into glucose in the liver before it can be metabolised (Jeukendrup, 2004). However, the values recorded at half-time and at the completion of the protocol were consistent for both drinks. This observation suggests that both glucose only and mixed carbohydrate drinks are effective at maintaining plasma glucose levels during soccer-specific exercise.

The rate of total carbohydrate oxidation was similar for both trials and may explain the similar levels of muscle glycogen and plasma glucose. The ingestion of mixed carbohydrate drinks compared with a drink containing an isoenergetic amount of glucose can increase exogenous carbohydrate oxidation in a thermo-neutral environment (Jentjens *et al.*, 2004) and in the heat (Jentjens *et al.*, 2006), although total carbohydrate oxidation is unaffected, as was observed in the present study. This observation is consistent with previous studies although carbohydrate was ingested at higher concentrations (1.5–2.4 $g \cdot min^{-1}$) (Jentjens *et al.*, 2004; Jentjens *et al.*, 2006) than in the present study.

In the present study weight loss, expressed as a percentage, and sweat loss were not significantly different between the two drinks conditions. These findings suggest that there was no significant difference in the amount of water absorbed from the intestine between the different types of drink. Previous studies (Davis *et al.*, 1990; Jentjens *et al.*, 2006) have indicated that fluid availability during exercise in the heat is lower with a glucose drink compared with a combined glucose and fructose drink or water. Shi *et al.* (1995) also demonstrated that the ingestion of a glucose and fructose solution resulted in greater water absorption than a glucose solution. In addition, Jentjens *et al.* (2006) reported greater (although not statistically significant) changes in plasma volume with glucose compared with a multi-carbohydrate drink and concluded that a glucose drink may be less effective for fluid replacement during exercise in the heat. During prolonged exercise in the heat, it has been suggested that fluid containing more than 2.5% carbohydrate inhibits fluid delivery (Costill and Saltin, 1974). However, Hawley *et al.* (1991) demonstrated that fluid containing as much as 15% carbohydrate was as effective as water in supporting thermoregulatory processes and performance during exercise in the heat. Therefore, it appears to be appropriate to consume drinks with a carbohydrate content of 6.6%, as used in the present study, when playing soccer in the heat.

In conclusion the ingestion of a solution containing glucose and fructose compared with an isoenergetic glucose solution did not significantly influence muscle glycogen utilization, the metabolic responses to soccer-specific exercise performed in the heat or high-intensity exercise capacity measured post-exercise. The results suggest that intestinal absorption does not limit the oxidation of exogenous carbohydrate during exercise of this nature in the heat.

Acknowledgements

GlaxoSmithKline is acknowledged for its contribution to this study.

References

Costill, D.L. and Saltin, B., 1974, Factors limiting gastric emptying during rest and exercise. *Journal of Applied Physiology,* **37,** pp. 679-683.

Coyle, E.F. and Coggan, A.R., 1984, Effectiveness of carbohydrate feeding in delaying fatigue during prolonged exercise. *Sports Medicine,* **1,** pp. 446-458.

Cunningham, D.A. and Faulkner, J.A., 1969, The effect of training on aerobic and anaerobic metabolism during a short exhaustive run. *Medicine and Science in Sports,* **1,** pp. 65-69.

Davis, J.M., Burgess, W.A., Slentz, C.A. and Bartoli, W.P., 1990, Fluid availability of sports drinks differing in carbohydrate type and concentration. *American Journal of Clinical Nutrition,* **51,** pp. 1054-1057.

Davis, J.M., Lamb, D.R., Pate, R.R., Slentz, C.A., Burgess, W.A. and Bartoli, W.P., 1988, Carbohydrate-electrolyte drinks - effects on endurance cycling in the heat. *American Journal of Clinical Nutrition,* **48,** pp. 1023-1030.

Drust, B., Reilly, T. and Cable, N.T., 2000, Physiological responses to laboratory-based soccer-specific intermittent and continuous exercise. *Journal of Sports Sciences,* **18,** pp. 885-892.

Frayn, K.N., 1983, Calculation of substrate oxidation rates in vivo from gaseous exchange. *Journal of Applied Physiology,* **55,** pp. 628-634.

Hawley, J.A., Dennis, S.C., Laidler, B.J., Bosch, A.N., Noakes, T.D. and Brouns, F., 1991, High rates of exogenous carbohydrate oxidation from starch ingested during prolonged exercise. *Journal of Applied Physiology,* **71,** pp. 1801-1806.

Jentjens, R.L.P.G., Underwood, K., Achten, J., Currell, K., Mann, C.H. and Jeukendrup, A.E., 2006, Exogenous carbohydrate oxidation rates are elevated after combined ingestion of glucose and fructose during exercise in the heat. *Journal of Applied Physiology,* **100,** pp. 807-816.

Jentjens, R.L.P.G., Achten, J. and Jeukendrup, A.E., 2004, High oxidation rates from combined carbohydrates ingested during exercise. *Medicine and Science in Sports and Exercise,* **36,** pp. 1551-1558.

Jeukendrup, A.E., 2004, Carbohydrate intake during exercise and performance. *Nutrition,* **20,** pp. 669-677.

Jeukendrup, A.E. and Jentjens, R., 2000, Oxidation of carbohydrate feedings during prolonged exercise - Current thoughts, guidelines and directions for future research. *Sports Medicine,* **29,** pp. 407-424.

Lowry, O.H. and Passonneau, J.V., 1972, *A Flexible System of Enzymatic Analysis.* (Academic Press: New York.)

Morris, J.G., Nevill, M.E., Boobis, L.H., Macdonald, I.A. and Williams, C., 2005, Muscle metabolism, temperature, and function during prolonged, intermittent, high-intensity running in air temperatures of 33 degrees and 17 degrees C. *International Journal of Sports Medicine,* **26,** pp. 805-814.

Morris, J.G., Nevill, M.E., Thompson, D., Collie, J. and Williams, C., 2003, The influence of a 6.5% carbohydrate-electrolyte solution on performance of prolonged intermittent high-intensity running at 30°C. *Journal of Sports Sciences,* **21,** pp. 371-381.

Morris, J.G., Nevill, M.E., Lakomy, H.K.A., Nicholas, C.W. and Williams, C.,

1998, Effect of hot environment on performance of prolonged, intermittent, high-intensity shuttle running. *Journal of Sports Sciences,* **16**, pp. 677-686.

Reilly, T. and Thomas, V., 1976, A motion analysis of work-rate in different positional roles in professional football match play. *Journal of Human Movement Studies,* **2**, pp. 87-97.

Sawka, M.N., Young, A.J., Francesconi, R.P., Muza, S.R. and Pandolf, K.B., 1985, Thermoregulatory and blood responses during exercise at graded hypohydration levels. *Journal of Applied Physiology,* **59**, pp. 1394-1401.

Shi, X., Summers, R.W., Schedl, H.P., Flanagan, S.W., Chang, R. and Gisolfi, C.V., 1995, Effects of carbohydrate type and concentration and solution osmolality on water absorption. *Medicine and Science in Sports and Exercise,* **27**, pp. 1607-1615.

Thomas, C., Plowman, S.A. and Looney, M.A., 2002, Reliability and validity of the anaerobic speed test and the anaerobic shuttle test for measuring anaerobic work capacity in soccer players. *Measurement in Physical Education and Exercise Science,* **6**, pp. 187-205.

Wallis, G.A., Rowlands, D.S., Shaw, C., Jentjens, R.L. and Jeukendrup, A.E., 2005, Oxidation of combined ingestion of maltodextrins and fructose during exercise. *Medicine and Science in Sports and Exercise,* **37**, pp. 426-432.

Part V
Psychology

CHAPTER TWENTY-TWO

Psychological preparation and development of players in Premiership football: Practical and theoretical perspectives

Mark Nesti and Martin Littlewood

Research Institute for Sport and Exercise Sciences, Liverpool John Moores University, Liverpool, UK

1. INTRODUCTION

We provide a series of practitioner reflections on engaging in applied sport psychology support in professional football (soccer). We draw specifically on our engagement in the world of professional football spanning over 7 years and involving 2 to 3 days delivery per week within the club setting. The aim is to illustrate *our* reality of providing sport psychology support in both Academy and 1st team professional environments. We conclude by offering some key issues for consideration in the training and development of applied sport psychologists. We feel that the material presented within this review is unique and challenges previous accounts in the literature. Of major significance, we argue that effective delivery is only possible where the psychologist is cognisant of the cultural demands and norms of the environment. These conditions influence the type of programme that can be offered in the club and the psychologist's approach to practice.

2. APPLYING SPORTS PSYCHOLOGY SUPPORT

Given the distinctive environment of professional soccer, we propose that an appropriate and well received sport psychology service is most likely when several features exist at a club. Some of these can be considered as essential to the success of any work to be done. Others may be viewed as desirable, in that their absence would not prevent work going ahead, but would enable it to progress more easily and effectively if they were in place. A number of researchers have highlighted how important confidentiality (Andersen, 2005), support from coaches (Gould *et al.*, 2001) and sports performers' attitudes towards psychology (Pain and Harwood, 2004) are to the successful delivery of sport psychology in team settings. However, there is no research specifically addressing these topics as they relate to Premiership soccer. This lack could be because sport psychologists and researchers working in soccer are unaware of the importance of such factors, although given their appearance in other team sports this seems unlikely. We propose that the reason issues of confidentiality and players' receptiveness towards psychological support are rarely mentioned is because of difficulties in gaining long-term sustained access to elite professional soccer teams.

In our experience, the most essential factor influencing the successful integration of sport psychology support into a soccer club is the attitude of managers to sport psychologists and their view on the importance of psychology in the game. There is little doubt that without the full and vigorous support of the manager very little can be achieved. It is also crucial that the manager sees psychological support as a long-term option rather than a quick fix solution (Corlett, 1996) and is prepared to accept that impact will be harder to measure directly when compared to other sports science disciplines. Where a manager expects to see immediate and clear results from psychological work with elite level professional footballers, it is likely to lead to considerable discomfort for the sport psychologist and significant frustrations for the manager. Within a Premiership Academy context, the sport psychologist may have an even greater task to convince the manager about the value and significance of his/her work. This can be due to misperceptions about the context, culture and working practices of the Academy. For example, sport psychologists working in the Academy may be seen as being too supportive to players during critical transition periods, when 1st team staff members feel mental toughness (e.g., Crust, 2007) should be developed more independently. This emphasises the importance of a shared philosophy and vision between the Academy and 1st team managers towards psychological development and preparation.

In our view, sport psychologists, especially when working with 1st team players, would be better employed to follow the advice of Corlett (1996) and Nesti (2004), which emphasises a primary focus on deeper, longer-term change. It is suggested that coaches (and often the players themselves) are best placed and qualified to conduct mental skills-related work to develop specific psychological elements such as confidence and concentration. However, sport psychologists in professional soccer must do more than acknowledge that coaches are often highly skilled in the delivery of mental skills. Our role is to support coaches in this work through education, rather than us having direct mental skills input with the players. In contrast, within an Academy context a systematic mental skills programme could be more easily delivered by the sports psychologist. This is because players are at a stage in their development where they are more receptive to educational work and interventions, and mental skills techniques may be perceived positively to assist their development and performance. This learning can be enhanced by support from coaches (Pain and Harwood, 2007) to help integrate skills into players' training and competitive experiences.

3. THE NATURE OF SPORTS PSYCHOLOGY WORK IN CLUBS

We would argue that sport psychology work in professional soccer could usefully be divided into organisational work and individually tailored support for players and coaches. Although there is a substantial body of research on group processes and team cohesion in sport (e.g., Carron *et al*., 2005) much of this does not deal in sufficient detail with some of the more specific issues prevalent in professional sports teams, or within Premiership soccer in particular. For example, the sport psychologist can provide a valuable role assisting with player liaison duties, including induction of players into the club, supporting players (and their families) to settle into a new area, and helping them to understand the team philosophy and overall approach to sports science provision, injury prevention and rehabilitation. In addition to these, Academy level issues can include integrating successfully into external educational programmes and living away from home for the first time. These issues are often not addressed in a systematic and coherent way in professional soccer, and if not dealt with properly can have an adverse psychological effect on the player. This is likely to be even more pronounced when the player is from another country and a very different culture. Given that from 1998 to 2005 there has been a significant decline in English-born professional Premier League players and a rapid increase in foreign-born nationals (Littlewood and Richardson, 2006), such work is likely of increasing importance.

The nature of this type of work with players led Gilbourne and Richardson (2006) to suggest that sport psychologists should provide a humanistic, caring and personal support service in professional soccer. We would subscribe to this notion and contend that in our experience, such an approach is likely to enhance players' psychological well-being and ultimately their performance. We believe that many of the performance difficulties new players have encountered in the clubs we have worked at relate to problems with broader life issues. Those problems are mainly associated with moving into a new environment and culture. In developing this idea, Nesti (2004) has proposed that sport psychologists should be educated and trained to deal with much more than mental skills training. This is because performance is often affected by factors like coach-athlete relationships, dealing with financial matters, unfamiliar work practices and adapting to new roles, which cannot be satisfactorily resolved through the use of goal setting, imagery or relaxation strategies.

Further tasks that could be couched under the heading of organisational psychology, yet fall within the role of the sport psychologist, include mentoring and developing other sports science staff, coaches and managers, devising codes of conduct, facilitating staff away days, writing job descriptions and clarifying roles. These types of activities are often carried out by occupational psychologists or human resource managers in other environments. In our experience this kind of work is often left undone, or may be picked up by other staff lacking the necessary expertise and time to carry them out effectively. The overarching aim here is to help the club create a culture of professionalism, dynamism and excellence, and ultimately to support players to perform optimally. We have seen sport psychologists attempt to engage in these activities, having recognised their importance to the team culture and environment. Often these individuals do not have all of the necessary experience or requisite skills to perform such tasks but do deliver them since this is expected by the club, and is something they know will impact positively on the psychological state of staff and players. Nesti (2004) has criticised the education and training programmes of sport psychologists for largely failing to prepare them to deal with this area of work. He pointed out that this often leads to misunderstanding and frustrations for both the club and the psychologist.

A role familiar to most sport psychologists would be the need to offer one-to-one support (to players) that is aimed directly at enhancing their performance (on the pitch). The approach we have taken in our work with Premiership footballers takes two forms. All 1st team professional players meet with the sport psychologist to complete a personality based psychometric test. The sports psychologist has been trained and is qualified to administer this psychometric test. This instrument has been used extensively in performance settings since it was originally validated in 1955. These data are made available to the player and the staff, and are used to help the staff to communicate more effectively with each player. The data are also used by the sport psychologist to help players make necessary changes. For us, this is about players achieving congruency between their real self and their self concept. These empirical data link very closely to humanistic and existential psychology principles around the concept of authenticity. The sport psychologist uses this information in a counselling dialogue with the player, where the focus is on getting them to reflect on this existential concept. This approach rests on the idea that we perform best when we are true to ourselves, and that crucially, we are often unable to do this because we do not really know who we are and what we stand for (Ravizza, 2002). This involves much more than reflective practice, which has been found to be a useful skill for many involved in sport (e.g., Knowles *et al.*, 2007).

It should be emphasised that the psychometric data are not used in team selection, although coaches have utilised the information at times to place players in different positions or help them assume new roles. Counselling sessions with players are aimed at helping them to confront the anxiety that accompanies learning and growth, and to accept that their path to greater performance and success will always involve sacrifices and difficult moments in their pursuit of excellence (Nesti, 2007). This balanced and real perspective, grounded in the psychology of existentialism and phenomenology, has been well received by 1st team Premiership players. It is not excessively optimistic or pessimistic, recognises that we have some measure of freedom to choose the kind of person we want to be, and seems closest to a common sense view of life. This last point may be one of the reasons why so few psychologists espouse this approach. The approach does not rely on using mental skills techniques in work with players, since this can interfere with the deeper, more personal engagement needed for real and lasting change. Within sport psychology literature some of those advocating this view (e.g., Corlett, 1996; Dale, 1996; Ravizza, 2002; and Nesti, 2004) have worked extensively in high level and professional sports where athletes usually possess excellent mental skills and are looking for something of depth and breadth. For example, Ravizza has pointed out that developing self-awareness may be more useful to the athlete, whilst both Corlett and Nesti have claimed that athletes should pursue self-knowledge and personal autonomy.

From an Academy rather than 1st team perspective, our one-to-one work with

Premiership players extends beyond the teaching of mental skills and deals with issues such as homesickness, parental and peer-related expectations, education, social demands and communication with coaches. These sessions take place on a more informal and ad hoc basis. They tend to be reactive and are aimed at meeting players' immediate needs as and when these arise. Geographically, the work extends to phone contact, private conversations at the club training ground, on coach journeys to competitive games and in the players' dining room. These 'contacts' are all aimed at dealing with specific issues and developing a trusting relationship between the sport psychologist and player. In this sense, our practice is informed by a humanistic and person centred agenda (Gilbourne and Richardson, 2005).

4. CONFIDENTIALITY AND TRUST

To carry out a programme of one-to-one sport psychology counselling meetings with Premiership players, there are two essential elements that must be in place. First, it is essential that these interactions are carried out confidentially. We have used a recording system, in common with other sports science and medical staff, which indicates if a player is meeting with the psychologist, whether it is an initial assessment or part of ongoing work, and the date and duration of the meeting. Beyond this basic level of information, we have given confidential reports from the meetings to the player and established agreement that the manager, coaches and other key staff members are unable to access any information about the sessions from the psychologist or the player. This allows the player to discuss any topic with the sport psychologist. Without this level of complete confidentiality, players would be very unlikely to discuss (as they have) such matters as their desire to leave the club, contract issues, relationships with the manager and other important topics that are likely to impact on their playing performance if they are ignored. Secondly, due to the very unique and idiosyncratic culture of elite level professional soccer, it is essential that the manager trusts the sport psychologist and believes that there is a real value in players being given the chance to talk about themselves in this way to a qualified individual. In our experience, there is added strength and value if the psychologist is seen as an outsider who is nevertheless an insider to some degree in that he or she is considered to be part of the team by the players and staff. This has a positive impact on the players, knowing that the sport psychologist is not involved for just a brief intervention and has the respect of the other staff while clearly being somewhat detached and independent from selection and the day-to-day running of the club. In addition, trust is built up by this arrangement which is arguably the most important factor governing the effectiveness of the sport psychologist in this environment.

5. RECOMMENDATIONS

The future training and education of sports psychologists intending to operate in this environment needs to reflect the demands discussed within this review. Most notably, we feel that programmes of training and supervision need to highlight the importance of the context and cultural demands associated with this environment. This development could be supported by providing greater access to placements and through providing conferences, workshops and seminars specifically addressing the particular demands associated with delivering sport psychology in professional soccer. This should result in an increase in levels of understanding and practitioner skills. These in turn, will directly impact on players' attitudes towards sports psychology support and the ability of the practitioner to meet their needs. We strongly believe that successful practice will involve much more than mental skills training. In helping sport psychologists to develop the skills and personal qualities to work in soccer more emphasis should be placed on acquiring knowledge and competence in counselling psychology. This may go some way towards convincing sport psychologists that there needs to be congruence between their personal beliefs and their approach to practice. Indeed staying true to oneself, one's own values and philosophy of practice may be the only way, and ultimately, the most effective approach to working as a sports psychologist in the exciting, rewarding and volatile world of Premiership soccer.

References

Andersen, M.B., 2005, "Yeah, I work with Beckham": issues of confidentiality, privacy and privilege   in sport psychology service delivery. In *Sport and Exercise Psychology Review*, **1** (The British Psychological Society), pp. 5-13.

Carron, A.V., Patterson, M.M. and Loughead, T.M., 2005, The influence of team norms on the cohesion self-reported performance relationship: a multi-level analysis. *Psychology of Sport and Exercise*, **6**, pp. 479-493.

Corlett, J., 1996, Sophistry, Socrates and sport psychology. *The Sport Psychologist*, **10**, pp. 84-94.

Crust, L., 2007, Mental toughness in sport: a review. *International Journal of Sport and Exercise Psychology*, **5**, pp. 270-290.

Dale, G. A., 1996, Existential phenomenology: emphasizing the experience of the athlete in sport psychology research. *The Sport Psychologist,* **10**, pp. 307-321.

Gilbourne, D. and Richardson, D., 2005, A practitioner-focussed approach to the provision of psychological support in soccer: Adopting action research themes and processes. *Journal of Sports Sciences,* **23**, pp. 651-658.

Gilbourne, D. and Richardson, D., 2006, Tales from the field: Personal reflections on the provision of psychological support in professional soccer. *Psychology of Sport and Exercise,* **7**, pp. 325-337.

Gould, D., Greenleaf, C., Guinan, D., Dieffenbach, K. and McCann, S., 2001, Pursuing performance excellence: Lessons learned from Olympic athletes and coaches. *Journal of Excellence*, **4**, pp. 21-43.

Knowles, Z., Gilbourne, D., Tomlinson, V. and Anderson, A., 2007, Reflections on the application of reflective practice for supervision in applied sport psychology. *The Sport Psychologist*, **21**, pp. 109-122.

Littlewood, M. and Richardson, D., 2006, Football labour migration: Player acquisition trends in elite level English professional football (1990/91-2004/05). 14th European Sport Management Congress, Nicosia, Cyprus.

Nesti, M. S., 2004, *Existential Psychology and Sport: Theory and Application*. (London: Routledge.)

Nesti, M.S., 2007, Persons and players. In *Sport and Spirituality: An Introduction*, edited by Parry, J., Nesti, M.S., Robinson, S. and Watson, N. (London: Routledge), pp. 7-21.

Pain, M. and Harwood, C., 2004, Knowledge and perceptions of sports psychology in English soccer. *Journal of Sports Sciences,* **22**, pp. 813-826.

Pain, M. and Harwood, C., 2007, The performance environment of the England youth soccer teams. *Journal of Sports Sciences*, **25**, pp. 1307-1324.

Ravizza, K., 2002, A philosophical construct: A framework for performance enhancement. *International Journal of Sport Psychology*, **33**, pp. 4-18.

CHAPTER TWENTY-THREE

A comparison of PETTLEP imagery, physical practice and their combination in the facilitation of non-dominant leg kicking accuracy

Jon Finn, Alastair Grills and Dan Bell

Carnegie Research Institute, Leeds Metropolitan University, Leeds, UK

1. INTRODUCTION

Developing expertise in soccer has been suggested to involve 10,000 hours of deliberate practise extended over 10 years (for reviews see Williams and Hodges, 2004). Traditionally, much of this deliberate practise has been completed in the form of physical practise. However, there are two obvious limitations to relying on physical practise: (i) the onset of muscle fatigue and (ii) the risk of overuse injuries. Therefore there may be value in identifying and refining alternative forms of deliberate practise to accompany and complement the physical form; one such alternative may be mental practice, which is a form of imagery. The aim of this study is to test the efficacy of an imagery intervention, a physical practise intervention and a combination of both interventions in the advancement of kicking accuracy in soccer players.

Imagery is a popular and diverse mental skill, and its use is advocated by sport psychologists (Driskell *et al.*, 1994; Munroe *et al.*, 2000). Cognitive specific imagery, or motor imagery, is one of imagery's five key applications (Paivio, 1985) and can be used to rehearse a specific motor skill, for example kicking a ball. To optimise the efficacy of motor imagery, the functional equivalence between the imagined and motor tasks should be maximised (Hall, 2001).

The term functional equivalence has been used to describe the extent to which the processes of motor imagery, overt motor behaviour and action observation share common neural structures (Keil *et al.*, 2000). For example, when an individual images, performs and observes a specific motor task, this procedure activates some similar neural structures, including the perietal cortex, the supplementary motor area, premotor cortex, prefrontal cortex, cerebellum, primary motor cortex and the spinal cord (Papaxanthis *et al.*, 2002). The PETTLEP model of motor imagery (Holmes and Collins, 2001) aims to functionalize the equivalence between imagery and overt motor behaviour, therefore attempting to maximise the activation of shared neural circuitry when imagining the performance of a motor task (Holmes, 2007).

Each letter of the acronym PETTLEP describes a component of physical practice that should be considered when performing motor imagery to enhance functional equivalence. For example, when imaging a motor task such as a penalty kick in a major cup final, PETTLEP suggests the following should be taken into consideration: *Physical*; imagery should contain movements which reflect those of motor preparation and execution in the physical skill, such as body position and stance in relation to a penalty kick. *Environmental*; elements such as crowd noise should be incorporated into the image. *Task*; the task being imaged should be specific to the performer in terms of how the task feels and looks when they are performing. *Timing*; the time taken to perform the imagery should reflect how long it takes to physically perform the penalty kick. *Learning*; performer-specific skill level should be given consideration - taking a penalty may be more automatic to skilled performers than novices, meaning that each may image the penalty kick differently. *Emotion*; athletes should generate emotional responses appropriate to the level of excitement or nervousness that may be induced by a penalty kick in an important event. *Perspective*; the optimal imagery perspective should be considered for the individual athletes, whether internal or external.

Several researchers have investigated and supported the efficacy of PETTLEP -based imagery (an adaptation of PETTLEP imagery) in facilitating the acquisition of sports specific motor skills (Smith and Collins, 2004; Ramsey *et al.*, 2007; Smith *et al.*, 2007). For example, Smith and Collins found no significant differences in the strength gains made during a finger strength intervention across the conditions of physical practice and adapted forms of PETTLEP-based imagery. Smith and colleagues found further support for the motor learning effects of PETTLEP-based imagery in hockey penalty and gymnastic beam tasks. Again motor improvements made in PETTLEP-based imagery conditions were statistically significant and very similar to improvements made in physical practise conditions. In comparison, traditional imagery interventions did not yield significant improvements from pre-test to post-test.

Further research has provided support for the combined effects of PETTLEP-based imagery and physical practice (Smith and Holmes, 2004; Smith *et al.*, 2006). Smith and Holmes combined different imagery modality conditions with physical practice on a golf putting task. Post-tests revealed that participants using combinations of physical practice and imagery modalities which were strongly aligned to PETTLEP-based imagery (video and audio modalities) improved significantly more than participants using combinations of physical practice and a more traditional written imagery modality. Smith and colleagues also found support for the combined effect of PETTLEP-based imagery and physical practice in a golf bunker task. Further they found that participants in the combined PETTLEP-based imagery and physical practice group improved significantly more than participants in physical practice only and PETTLEP-based imagery-only conditions.

The aim in the current investigation was to increase understanding of the learning effects PETTLEP-based imagery might have when combined with physical practice in a pragmatic soccer coaching context. Specifically, we investigated the effect of PETTLEP-based imagery, physical practice and a combination of the two on the kicking accuracy of the non-dominant leg in soccer players.

2. METHODS

2.1 Participants

Twenty male university soccer players participated in the present study (mean age = 21.1 years, SD ± 0.72 years). Participants provided informed consent before the study commenced. Testing complied with the ethical guidelines of Leeds Metropolitan University. Participants were randomly assigned to one of four groups: PETTLEP-based imagery; physical practice; combined group (physical practice and PETTLEP-based imagery); control group.

2.2 Measures

Participants completed the Movement Imagery Questionnaire – Revised (MIQ-R) (Hall and Martin, 1997), which is an eight-item inventory measuring an individual's ability to perform kinaesthetic and visual imagery. Four items assess kinaesthetic imagery ability; e.g. Item 1: attempt to feel yourself making the movement just performed without actually doing it, with four items for visual imagery ability e.g. Item 2: attempt to see yourself making the movement just performed with as clear and vivid a visual image as possible. In line with previous research and recommendations (see Hall, 1998; Smith and Collins, 2004; Smith *et al.*, 2007) potential participants were excluded if they scored lower than 16 on either the kinaesthetic or visual subscales. Furthermore, participants were retained in the study only if they had a preference for external visual imagery, as in the intervention phase only realistic external images of the participants taking penalty kicks (external visual primers) could be generated by the researchers. These visual primers were key features in two of the four intervention groups.

2.3 Equipment

In line with previous soccer specific research (see Finnoff *et al.*, 2002) two identical soccer-specific targets measuring 1210 mm x 1210 mm were built for this study. The targets were attached to the framework of a five-a-side soccer goal (3800 mm x 1300 mm) for the purposes of ecological validity.

Targets were placed in adjacent corners of the goal (see Figure 1). The targets were coated in plain white paper with sheets of A4 carbon paper stapled on the exterior. Carbon paper was used to mark the connection between the ball and the target.

Figure 1. Targets for measuring kicking accuracy.

Two Sony Mini DV Handy Cam video cameras were used to record participants who had been randomly assigned to either the PETTLEP imagery or the combined group. This provided an imagery source for the intervention. These clips were edited and saved to DVDs using Dartfish software. Edited video footage provided visual primers to control for imagery perspective, agency and modality. All visual primers were of an external imagery nature. In accordance with Hardy and Callow (1999), external images were captured by a camera placed 8 m away from the penalty spot at a 60° angle from the ball, and a second camera placed 4 m away from the penalty spot at a 160° angle (see Figure 2).

2.4 Procedure

The criterion task used was a soccer penalty kick. Each penalty kick was taken from the regulation position - 12 yards (11 metres) from the midpoint between the goalposts and equidistant to them (Federational Internationale de Football Association, 2007). Targets were placed within both corners of the goal frame. A ball of standard size 5 and inflation (pressure: 0.8 bar) was used for both pre-test and post-test condition. Participants were advised that accuracy was being assessed in relation to how close they could get the ball to the centre of the target and power was of minor significance, although they should be striking to score past a goalkeeper. No goalkeeper was present, as variability in participant responses and goalkeeper behaviour may have confounded results (Smith *et al.*, 2001).

Participants were instructed to adopt a three-step approach prior to striking the ball with the non-dominant foot. In the pre-test players kicked the ball at the right-hand target five times and then the left-hand target five times. This same procedure was conducted for the post-test. In accordance with Smith *et al.* (2007), the participants were permitted five practise shots with their non-dominant foot at both targets before pre-tests and post-tests.

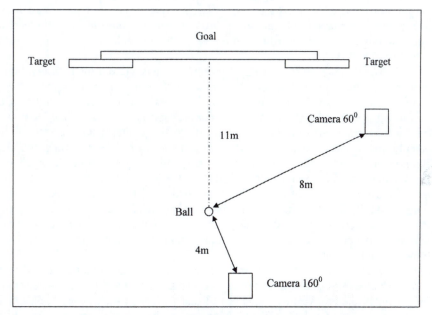

Figure 2. Camera angles in relation to penalty taker and the goal.

2.5 Scoring

Scores were based on measuring the distance between the centre point of the target and the centre point of the mark left by the ball. Scores that deviated away from the centre point of the target were considered to be absolute error (i.e., the distance between the mark left by the carbon paper and the target centre point). The closer the ball was to the centre point of the target, the lower the absolute error. To enhance reliability, the same individual scored all kicks. An arbitrary score of 1000 mm was assigned for attempts missing the target entirely. This value was thought to be a valid and reliable marker of missed attempts. Intervention effectiveness was measured by comparing the difference between pre-test and post-test mean kicking accuracy data.

2.6 Intervention

Following the pre-test data collection, participants were randomly assigned, in relation to the order that they were tested, into one of the four intervention groups, with each group containing five participants. The researchers provided participants within the PETTLEP imagery and combined groups with individual specific edited video footage that contained observational primers of their performance to control for imagery perspective, agency and modality. All participants were provided with diaries subsequent to pre-tests. The diaries were used to record structured qualitative information relating to specific questions concerning their specific interventions and it was thought that they would help to enhance adherence (Smith and Holmes, 2004).

PETTLEP group: The PETTLEP group members viewed their observational primer once, before then imaging themselves performing the penalty kick. They were instructed to do this in a quiet place of their convenience. Participants performed their imagery in the appropriate sportswear, including footwear (physical component), and were instructed to image themselves executing the penalty kick perfectly with their non-dominant foot in real time (timing and task components) and from an external perspective (perspective component). Participants imagined performing the kick ten times, with five kicks aimed to the left of the target and five kicks aimed the to the right. This was repeated three times a week, resulting in 180 imaged penalty kicks over a six-week period. Participants were encouraged to incorporate any of their usual pre-kick routines into their imagery to replicate the real penalty-taking scenario as much as possible (physical component).

Physical practice: Physical practice participants were required to take 10 penalty kicks three times a week for six weeks with their non-dominant foot. Participants undertook their practice on an astroturf pitch, using goals and imaginary targets of a similar size to those used in the pre-tests and post-tests. Five shots were taken at the right imaginary target and five at the left imaginary target. The physical practice group completed 180 penalty kicks over a six-week period.

2.6.1 Combined group

The combined group completed one physical practice session and two PETTLEP sessions per week, as described above. The combined group completed 60 physical penalty kicks and 120 imaged penalty kicks over a six-week period.

2.6.2 Control group

Control participants were required to read literature related to soccer, specifically match reports located on the Sky Sports website; questions within the diary relating to the literature were then answered. The control participants were required to spend 10-15 minutes engaging in their reading and answering the questions, matching the time that the other three groups spent on their intervention activity. This procedure was completed three times a week over a six-week period.

2.6.3 General intervention procedures

All participants across all intervention groups were instructed to adopt precisely the same process for each session. After each session participants completed the questions in their respective diaries. Questions within the diaries were related to their allocated intervention activity. Whilst engaging in this study, participants were instructed to carry out their normal soccer practices and behaviour. This was believed to be an approximately equivalent level of soccer activity across all participants. It was explained to all participants that they should not engage in any physical practice that simulate the current study, unless this was a requirement of their intervention. After the six-week period all diaries were collected and all participants completed the post-tests.

2.7 Statistical analysis

Data were screened via t-tests to confirm assumptions of normality. Preference for these conservative measures was made given the likelihood of low recruitment numbers. The data conformed to assumptions of ANOVA testing.

3. RESULTS

3.1 Performance data

Statistical analysis using an ANOVA revealed no significant treatment effects. However, in a second stage of analysis, which excluded target misses (73 trials), an ANOVA showed a treatment effect where the combination intervention was significantly more effective than the control intervention ($F_{3,113}$ = 3.97, P<0.01). The average intervention advantage was 251.29 (SD ± 77.03) mm. No other treatment effects were found according to measured variables.

Descriptive analysis of the mean kicking accuracy (Figure 3) revealed a 24.2% (pre-713 mm vs post-540 mm) increase in the combined group, a 21.74% increase (pre-730 mm vs post-571 mm) in the physical group, a 18.6% (pre-test 700 mm vs post-test 570 mm) increase in the PETTLEP group and a 3.3% increase in the control group.

3.2 Self-report data

Due to injury following pre-testing, one participant from the PETTLEP group was unable to complete the intervention. Therefore PETTLEP group data relied on information gained from four participants. Manipulation checks were carried out to ensure participants had adhered to their intervention programmes by inspecting their diaries. Adherence to the study was as follows: PETTLEP group 100%; physical practice group 88%; combined group 88%; control group 100%.

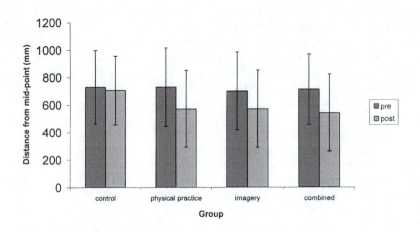

Figure 3. Group means and standard deviations for pre-tests and post- tests.

4. DISCUSSION

We examined the effects of PETTLEP imagery, physical practice, and a combination of the two on the kicking accuracy of the non-dominant leg in recreational university soccer players. Results supported previous researchers (Smith *et al.*, 2006) in demonstrating that a combination of PETTLEP-based imagery and physical practice can have a significant impact on the acquisition of sport specific skills. However, results from this study did not support researchers (Smith and Collins, 2004; Smith and Holmes, 2004; Smith *et al.*, 2007) who have reported PETTLEP-based imagery only (i.e., not in combination with physical practice) to have a significant impact on sport-specific skill acquisition. Furthermore, these results showed no significant impact on the learning of sport specific skills through physical practice alone.

In the present study the combined intervention did not allocate practice resources evenly between physical practice and imagery. As the combined group participants completed 60 physical and 120 imagined penalty kicks the findings, therefore, lend some support to the efficacy of PETTLEP-based imagery engaging shared neural circuitry and facilitating the acquisition of a sports-specific motor task (Holmes and Collins, 2001).

There are a number of possible explanations for the success of the combined group in significantly improving kicking accuracy while the physical practice and PETTLEP-based imagery groups did not. First, participants in the combined group may have been more motivated by the variety in their intervention, which may have been amplified by the duration of the trial. Groups completing physical practice or PETTLEP-based imagery only may have experienced boredom over the period of their intervention which may have reduced the mental effort that they invested into each individual practice trial. These issues were not measured in our manipulation checks and therefore we do not have direct evidence to support this explanation. Second, participants were instructed to undertake their practice schedules in a highly blocked format. Motor control research would suggest that random, serial or variable practice would increase contextual interference and enhance the benefits of practice for performers at the skill level depicted by university-level soccer players. Further, this may support the significant learning effects observed in the combination group due to the mixture of practice methods used.

We provided clearly written imagery instructions and observational primers to control for imagery perspective, agency and modality. These were provided in an attempt to maximise shared neutral circuitry in the PETTLEP-based imagery intervention. However, since kicking performance was not significantly improved in the PETTLEP only intervention group, procedures should be reconsidered in order to increase shared neural circuitry further. In future, researchers might, for example, allow participants access into the data collection hall to increase the environmental stimulus in the imagery. Furthermore, the perspective component of the PETTLEP model may be advanced by including 360 degree video clip angles of participants kicking the ball to replace the potentially limiting external perspective used in the present study.

There are several further limitations to the present study, which should subsequently be addressed. First, the target used to measure kicking accuracy lacked the sensitivity to distinguish penalties directed more than 1000 mm from the target centre. Second, there was only limited capacity to control participants' imagery ability. Although potential participants were excluded from the study if they scored lower than 16 on either the kinaesthetic or visual subscales of the MIQ-R, random allocation to intervention groups might have been conducted according to imagery ability to reduce imagery ability bias in any one group. Moreover, participants' imagery ability could have been re-tested during post-tests to monitor any changes in imagery skill. Third, although participants were all university-level players and randomly assigned to intervention groups, the researchers could have optimised the group allocation process. One such option could have been to rank participants by pre-test scores and use these ranks to determine group allocation, so helping to balance the distribution of skills between treatment groups. However, there is a possibility that with such a small sample size a ranking system may have falsely smoothed raw scores and therefore given a false sense of balance to the groups. Fourth, this small sample size resulted in low statistical power and this should be considered in light of statistically significant increases in kicking accuracy in the combined group. Finally, clear manipulation checks specifically measuring the self-reported mental effort invested in every single intervention trial would have given the researchers a better understanding of each participant's trial by trial mental engagement across interventions.

5. CONCLUSION

The present study provides evidence that PETTLEP-based imagery can be used effectively in combination with physical practice to enhance soccer specific motor skills. The limitations of this study may have restricted our capacity to illuminate the potential of PETTLEP in improving performance in closed-skill tasks like penalty taking. However, coaches should consider PETTLEP-based imagery to be another tool in their armoury as our descriptive data cautiously suggested that PETTLEP-based imagery did have a positive effect on participants' kicking ability. Most players from junior to elite level will openly admit that they can improve their non-dominant kicking leg. Conversely, attempts to develop this area are often inhibited by players avoiding using their non-dominant leg in training to save embarrassment, or because players are often unwilling to engage in extra physical practice with the non-dominant leg. PETTLEP-based imagery could provide an effective and novel tool to help coaches impact on this area of performance and go some way to optimising player potential.

References

Driskell, J.E., Copper, C. and Moran, A., 1994, Does mental practice enhance performance? *Journal of Applied Psychology*, **79**, pp. 481-492.

Fédération Internationale de Football Association, 2006, Laws of the game 2006 * (Online). Available from http://www.fifa.com/en/regulations (Cited 24 April 2007).

Finnoff, J., Newcomer, K., and Laskowski, E., 2002, A valid and reliable method for measuring the kicking accuracy of soccer players, *Journal of Science and Medicine in Sport*, **5**, pp. 348-353.

Hall, C., 2001, Imagery in sport and exercise. In *Handbook of Sport Psychology*, edited by Singer, R., Hausenblas, H. and Janelle, C. (New York: Wiley), pp. 529-549.

Hall, C.R., 1998, Measuring imagery abilities and imagery use. In *Advances in Sport and Exercise Psychology Measurement*, edited by Duda, J.L. (Morgantown: Fitness Information Technology), pp. 165-172.

Hall, C.R. and Martin, K.A., 1997, Measuring movement imagery abilities: A revision of the Movement Imagery Questionnaire. *Journal of Mental Imagery*, **21**, pp. 143-154.

Hardy, L. and Callow, N., 1999, Efficacy of external and internal visual imagery perspectives for the enhancement of performance on tasks in which form is important, *Journal of Sport and Exercise Psychology*, **21**, pp. 95-112.

Holmes, P.S., 2007, Integrating Imagery, Observation and Neuropsychology. In *Proceedings of the 12th European Congress of Sport Psychology*, edited by Theodorakis, Y., Goudas, M. and Papaioannou, A. (Halkidiki, Greece), pp. 140.

Holmes, P.S. and Collins, D.J., 2001, The PETTLEP approach to motor imagery: A functional equivalence model for sport psychologists. *Journal of Applied Sport Psychology*, **13**, pp. 60-83.

Keil, D., Holmes, P., Bennett, S., Davids, K. and Smith, N., 2000, Theory and practice in sport psychology and motor behaviour needs to be constrained by integrative modelling of brain and behaviour. *Journal of Sport Sciences*, **18**, pp. 433-443.

Munroe, K.J., Giacobbi, P.R., Jr., Hall, C. and Weinberg, R., 2000, The four Ws of imagery use: where, when, why, and what. *The Sport Psychologist.* **14**, pp. 119-37.

Paivio, A., 1985, Cognitive and motivational functions of imagery in human performance. *Canadian Journal of Applied Sport Sciences*, **10**, pp. 22S-28S.

Papaxanthis, C., Schieppati, M., Gentili, R. and Pozzo, T., 2002, Imagined and actual arm movements have similar duration when performed under different conditions of direction and mass. *Experimental Brain Research,* **143**, pp. 447-452.

Ramsey, R., Cumming, J., Brunning, C., Williams, S., 2007, A PETTLEP based imagery intervention with university soccer players. *Journal of Sport and Exercise Psychology*, **29**, pp. S196.

Smith, D. and Collins, D., 2004, Mental practice, motor performance and the late CNV. *Journal of Sport and Exercise*, **26**, pp. 412-426.

Smith, D. and Holmes, P., 2004, The effect of imagery modality on golf putting performance. *Journal of Sport and Exercise Psychology*, **26**, pp. 385-395.

Smith, D., Holmes, P., Whitmore, L., Collins, D., and Davenport, T., 2001, The effect of theoretically-based imagery scripts on hockey penalty flick performance. *Journal of Sport Behaviour*, **24**, pp. 408-419.

Smith, D., Wright, C. and Cantwell, C., 2006, Beating the bunker: PETTLEP imagery and golf bunker shot performance. In *Proceedings of the British Association of Sport and Exercise Sciences Annual Conference, Wolverhampton*, edited by Lane, A. (Wolverhampton, England), p. 12.

Smith, D., Wright, C., Allsopp, A. and Westhead, H., 2007, It's all in the mind: PETTLEP-based imagery and sports performance. *Journal of Applied Sport Psychology*, **19**, pp. 80-92.

Williams, A.M. and Hodges, N.J., 2004, *Skill Acquisition in Sport: Research, Theory and Practice* (London: Routledge).

CHAPTER TWENTY-FOUR

Coping strategies of injured semi-professional soccer players

Fraser Carson

Department of Sport and Physical Activity, Edge Hill University, Ormskirk, UK

1. INTRODUCTION

The majority of research in the psychology of injury rehabilitation has supported Lazarus and Folkman's (1984) view of the dynamic nature of coping, involving a range of differing coping strategies that include adaptive and recursive transactions between the environment and personal variables where an individual makes an appraisal of the situation (Lazarus, 1999). Kowalski and Crocker (2001) separated coping into three higher-order dimensions. Problem-focused coping relates to concerted efforts to manage a stressful situation (e.g., goal-setting, planning) whereas emotion-focused coping is aimed at controlling emotional reactions (e.g., deep breathing, visualization) (Lazarus, 1999). The third dimension, avoidance coping, is concerned with activities or cognitive changes to avoid the situation via distraction (e.g., blocking, cognitive distancing) or social diversion (e.g., walking away, removing oneself from the situation) (Endler and Parker, 2000).

Despite the "bad reputation" (Stanton and Franz, 1999) of emotion-focused coping both problem-focused and emotion-focused coping strategies have been identified as adaptive to injured athletes (Udry, 1997). Kerr and Miller (2001), in line with Folkman's (1991; 1992) goodness of fit model, proposed that when an athlete has full control over rehabilitation, problem-focused coping strategies are preferred; in contrast, athletes who are unable to change the situation are more suited to make use of emotion-focused coping strategies. On the other hand, Albinson and Petrie (2003) suggested that problem-focused coping strategies may become insufficient for athletes rehabilitating from injury when mood disturbance is increased, whereas Johnston and Carroll (1998) suggested that problem-focused coping results in better adherence in rehabilitation following injury.

The aim of this study was to investigate the coping strategies utilized by soccer players rehabilitating from anterior cruciate ligament (ACL) surgery, in order to gain further understanding of the "unique perceptions and perspectives" experienced by injured players (Shelley, 1999). An additional aim was to provide those in regular contact with injured players, with further knowledge of the psychological processes experienced.

2. METHODS

Five semi-professional male soccer players (age range 18-34 years), who had completed their full rehabilitation following anterior cruciate ligament (ACL) surgery and returned to competition within two months prior to data collection, participated. A mixed methodological (qualitative dominant – quantitative less dominant) approach was employed across three distinct phases of the rehabilitation process; (a) Non-Participation Phase – detailing initial injury and surgery; (b) Limited-Participation Phase – detailing the physical rehabilitation; (c) Return-to-Play Phase – detailing the final training sessions before returning to competition and the first three games back in full competition. Methods included a semi-structured retrospective interview, designed in accordance with the vast amount of current literature, and completion of the 'Coping with Health, Injuries and Problems (CHIP)' inventory (Endler and Parker, 2000). Each interview was transcribed verbatim and a hierarchal content analysis conducted (Patton, 1990). The CHIP inventory provides a score for four subscales related to coping (Distraction coping; Palliative coping; Instrumental coping; Emotional preoccupation), which allowed for triangulation of utilized coping strategies discussed during the interviews.

3. RESULTS

Content analysis identified 15 higher-order themes that were coalesced into general dimensions of Problem-focused coping, Emotion-focused coping, or Avoidance coping for each of the three phases. The non-participation provided the higher-order themes: Information Gathering, Focus on Physical Fitness, and Goal Setting (Problem-focused coping); Emotional Control (Emotion-focused coping); and Denial (Avoidance coping). The limited-participation phase was composed of the higher-order themes: Focus on Physical Fitness (Problem-focused coping); Social Support, Confidence Building, and Relaxation Techniques (Emotion-focused coping); and Distraction (Avoidance coping). The return-to-play phase identified the higher-order themes: Goal Setting and Focus on Involvement (Problem-focused coping); Confidence Building and Social Support (Emotion-focused coping); and Fear of Re-injury (Avoidance coping). (See Figures 1-3.) No statistical analysis was conducted on the CHIP data; rather this information was utilized to build trustworthiness for the identified themes (Creswell, 2003). The results are considered according to the three identified phases.

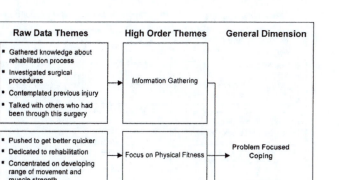

Figure 1. Non-Participation Phase.

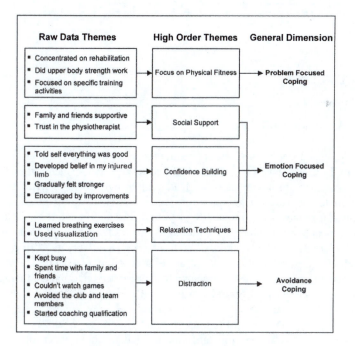

Figure 2. Limited Participation Phase.

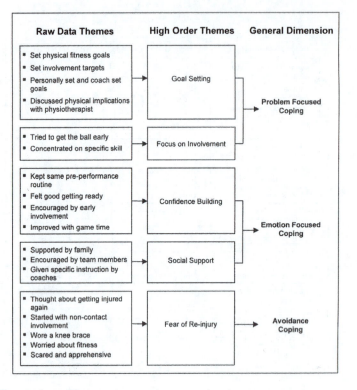

Figure 3. Return-to-play.

4. DISCUSSION

4.1 Non-Participation Phase

Problem-focused coping was most frequently used during this phase. Following initial ACL injury medical attention is required and as such problem-focused coping strategies are advantageous (Kerr and Miller, 2001). Particular emphasis was given to the need to gather information related to both the injury and the rehabilitation process. Comments by players included, "I tried to find out as much information as possible about the injury and surgery", and "I did loads of research on the internet about the injury". The CHIP data for the "find out more information" question averaged at 4.4 out of a possible 5. Having specific targets, focusing on the physical fitness aspects also appears to encourage adherence to the rehabilitation process (Evans and Hardy, 2002). Emotional control was regained by spending time with others and keeping busy. Having a good support structure can also facilitate the rehabilitation environment (Niven, 2007). Although no direct measure was taken, this appears to reduce the impact of anxiety and depression. The use of denial can have a debilitating effect, particularly at this level of performance where medical support may not be constantly accessible ("There wasn't a proper physio or anything like that to tell me how injured I was, so I just did what I thought I should"). Early diagnosis of the severity of the injury could improve acceptance and as such adherence (Grove and Gordon, 1995).

4.2 Limited Participation Phase

Autonomy in the rehabilitation programme was developed through problem-focused coping and should be encouraged (Levy *et al.*, 2008), particularly when the player is not in regular contact with the rehabilitation provider. The benefits of autonomy have only recently been discussed within the sports injury rehabilitation setting, but initial results have been positive (Podlog and Eklund, 2006; 2007). Developing confidence within both the injured knee and the physical body was highlighted as essential in the present study ("I felt stronger after every session and it gave me the belief that I was going to get back soon"). Performance accomplishment was essential to this process, with all players benefiting from progressions in the rehabilitation process. Social support and relaxation were also employed to assist controlling emotions (Albinson and Petrie, 2003). Kerr and Miller (2001) noted "emotion-focused coping strategies can help reduce distress and maintain optimism". This phase of the rehabilitation appears to indicate benefits with distraction coping, with CHIP scores on the distraction subscale being above the 68th percentile. Although the majority of literature recognizes the debilitating effect of avoidance (Kim and Duda, 2003), the present results suggest an advantage to reducing distress. One player commented, "I couldn't watch any of it. My team, the stuff on television, none of it. I just had to turn it off." However, further analysis needs to be conducted related to both cognitive and behavioural avoidance coping within sports injury settings.

4.3 Return-to-Play Phase

Again problem-focused coping strategies were utilized to gain autonomy and control during this phase of the rehabilitation (Podlog and Eklund, 2006). Specific goals assisted in focusing attention on personal involvement and functioned to encourage management of the situation ("I set myself targets to achieve, like playing for a certain length of time or making so many tackles"). These goals may need to become progressively more difficult over the duration of this phase, with specific emphasis on the use of process and performance goals (Evan and Hardy, 2002). Emotions were principally controlled by following previous performance routines (Jackson and Baker, 2001) and by encouragement from other team members (Magyer and Duda, 2000). Again both problem-focused and emotion-focused coping strategies combined to assist each player to control the stressful situation (Kaplan, 1996). There appears to be a cyclical relationship between the two ("The more I got involved, the more confident I became. The more confident I became the more challenging my goals became"); however, much further investigation is required. More research is necessary but contact and non-contact causes of the injury appear to dictate the level of fear associated with re-injury. These fears, including both injury and fitness worries (Tracey, 2003), appear to moderate the use of avoidance. Players noted that they delayed their return to competition and were inhibited in their performance when they eventually returned due to fears of re-injury.

5. CONCLUSION

Those involved with the rehabilitation of semi-professional soccer players may benefit from developing an autonomy-supportive environment, encouraging the players to take some control of their own rehabilitation. Utilizing problem-focused coping strategies and allowing them an increased involvement within the decision-making process can help the player to develop this autonomy and therefore reduce any negative consequences, such as non-adherence. Providing social support and encouragement appears to decrease the chances of depression affecting the player concerned. Avoidance coping may be beneficial to reduce the impact of negative emotions associated with withdrawal from the sport; however, care needs to be taken to avoid denial of the initial injury and fear of re-injury when returning to competition.

References

Albinson, C.B. and Petrie, T.A., 2003, Cognitive appraisals, stress and coping: Preinjury and postinjury factors influencing psychological adjustment to sport injury. *Journal of Sport Rehabilitation,* **12**, pp. 306-322.

Creswell, J.W., 2003, *Research Design: Qualitative, Quantitative and Mixed Methods Approaches* (Thousand Oaks, CA: Sage).

Endler, N.S. and Parker, J.D.A., 2000, *Coping with Health Injuries and Problems (CHIP)* (New York, NY: MHS).

Evans, L. and Hardy, L., 2002, Injury rehabilitation: A goal-setting intervention study. *Research Quarterly for Exercise and Sport,* **73**, pp. 310-319.

Folkman, S., 1991, Coping across the lifespan: Theoretical issues. In *Life-Span Developmental Psychology: Perspectives on Stress and Coping*, edited by Cummings, E. M., Greene, A. L. and Karraker, K. H. (Hillsdale, NJ: Lawrence Erlbaum Associates), pp. 3-19.

Folkman, S., 1992, Making the case for coping. In *Personal Coping: Theory, Research and Application*, edited by Carpenter, B.N. (Westport, CT: Praeger), pp. 31-46.

Grove, J.R. and Gordon, A.M.D., 1995, The psychological aspects of injury in sport. In *Science and Medicine in Sport*, 2nd ed., edited by Bloomfield, J., Fricker, P.A. and Fitch, K.D. (Cambridge, UK: Blackwell), pp. 194-205.

Jackson, R. and Baker, J., 2001, Routines, rituals and rugby: Case study of a world class goal kicker. *The Sport Psychologist*, **15**, pp. 48-65.

Johnston, L.H. and Carroll, D., 1998, The context of emotional responses to athletic injury: A qualitative analysis. *Journal of Sport Rehabilitation,* **7**, pp. 206-220.

Kaplan, H., 1996, *Psychosocial Stress: Perspectives on Structure Theory, Life Course and Methods.* (New York, NY: Academic Press.)

Kerr, G and Miller, P., 2001, Coping strategies. In *Coping with Sports Injuries: Psychological Strategies for Rehabilitation,* edited by Crossman, J. (Oxford, UK: Oxford University Press), pp. 83-102.

Kim, M.S. and Duda, J.L., 2003, The coping process: Cognitive appraisals of stress, coping strategies, and coping effectiveness. *The Sport Psychologist,* **17**, pp. 406-425.

Kowalski, K.C. and Crocker, P.R.E., 2001, The development and validation of the Coping Function Questionnaire for adolescents in sport. *Journal of Sport and Exercise Psychology,* **23**, pp. 136-155.

Lazarus, R.S., 1999, *Stress and Emotion: A New Synthesis.* (New York, NY: Springer.)

Lazarus, R.S. and Folkman, S., 1984, *Stress, Appraisal and Coping.* (New York, NY: Springer.)

Levy, A.R., Polman, R.C.J. and Borkoles, E., 2008, The role of perceived autonomy support and age upon rehabilitation adherence in sport. *Rehabilitation Psychology,* **53**, pp. 224-230.

Magyer, T.M. and Duda, J.L., 2000, Confidence restoration following athletic injury. *The Sport Psychologist,* **14**, 372-390.

Niven, A., 2007, Rehabilitation adherence in sport injury: Sport physiotherapists' perceptions. *Journal of Sport Rehabilitation,* **16**, pp. 93-110.

Patton, M.Q., 1990, *Qualitative Evaluation and Research Methods,* 2nd ed. (New Park, CA: Sage).

Podlog, L. and Eklund, R.C., 2006, A longitudinal investigation of competitive athletes' return to sport following serious injury. *Journal of Applied Sport Psychology,* **18**, pp. 44-68.

Podlog, L. and Eklund, R.C., 2007, Professional coaches' perspectives on the return to sport following serious injury. *Journal of Applied Sport Psychology,* **19**, pp. 207-226.

Shelley, G.A., 1999, Using qualitative case analysis in the study of athletic injury: A model for implementation. In *Psychological Bases of Sport Injuries* edited by Pargman, D. (Morgantown, WV: Fitness Information Technology), pp. 305-319.

Stanton, A.L. and Franz, A., 1999, Focusing on emotion: An adaptive coping strategy? In *Coping: The Psychology That Works,* edited by Synder, C.R. (New York, NY: Oxford University Press), pp. 90-118.

Tracey, J., 2003, The emotional response to injury and the rehabilitation process. *Journal of Applied Sport Psychology,* **15**, 297-293.

Udry, E., 1997, Coping and social support among injured athletes following surgery. *Journal of Sport and Exercise Psychology,* **19**, 71-90.

Part VI

Coaching

Innovative development in coach education: a foundation degree in football coaching and performance

Paul Rainer and Dave Adams

Division of Sport Health and Exercise, University of Glamorgan, Pontypridd, Wales

1. INTRODUCTION

A key factor in emerging sports coaching literature is the conceptualisation of coaching practice as a complex human interaction in a highly contextualised "difficult" set of circumstances. Consequently, Lyle (2002) has suggested that the recruitment of coaches is a more complex process than ensuring there is a pool of coaches available with an appropriate education and training. Currently, recruitment pathways into coaching are haphazard; there are enduring issues in the status and social standing of coaches; and coach education is characterised by individualised and ad-hoc learning pathways (Nelson and Cushion, 2006). Central to this process are the National Governing Body (NGB) awards. However, contemporary research would suggest that coach education courses are not preparing coaches for this role and the literature is unequivocal in adopting the position that coach education to date is perceived to lack relevance (Abrahams and Collins, 1998; Cushion et al., 2003).

It has been recognised that if we are to develop imaginative, dynamic and thought-provoking coaches we must widen the search beyond the content knowledge that has traditionally informed the coach education process (Cushion et al., 2003). Jones (2000) has suggested that traditional coach education courses have not developed the necessary intellectual and practical competencies, namely independent and creative thinking skills, which can only be developed through experiential learning. There remains a strong case that in its current form coach education in soccer fails to inform and influence practice (Cushion et al., 2003). This would suggest that trainee coaches need to be provided with supportive environments that allow them to apply theory to practice, in a range of contexts, to enable coaches to make mistakes, correct them, reflect on them and more importantly learn from them (Knowles et al., 2001).

The creation of the UK Coaching Framework (Sports Coach UK) has promoted the vision of coaching as a professionally regulated vocation. Further, delivery of a long-term coaching legacy and a world-leading UK coaching system are envisaged in the Framework by 2016 (Sports Coach UK, 2008), thus aiming to professionalise sports coaching. The coaching Framework (Sports Coach UK,

2008) clearly highlights the instrumental role that coaches will play in lifelong physical activity of young children and international success. The framework recognises the need for National Governing Bodies to align their coaching frameworks with Higher Education (HE) and some "joined-up thinking" in the delivery of coach education. Traditionally NGBs and HE have operated independently with regards to coach education, with little appreciation of the need to combine academic endeavour, certification and vocational experience. On consultation with key employers within the sport and recreation industry, Skills Active (2006) reported "the combination of Level 2 NGB coach and Level 4 knowledge would be extremely beneficial for sport", which suggests an integrated approach is required. It is clear from these proposed policies that coaches will play a vital role in developing and increasing participation and performance in sport at all levels. For this to happen it appears imperative that coach education programmes adhere to the concerns expressed by researchers and place emphasis on the development of tacit knowledge through experiential learning and reflection (Nelson and Cushion, 2006).

In attempts to address issues concerning the education of coaches, and in support of recommendations by Sports Coach UK, the University of Glamorgan has developed a Foundation Degree (FnD) in Football Coaching and Performance. Although only in its third year, the programme is proving to be highly successful with 100% of students to date passing both academic and vocational aspects of the course. This FnD programme has been developed in conjunction with key stakeholders in the industry (e.g., Cardiff City FC; Swansea City FC) and provides the opportunity for students to acquire their Football Association of Wales (FAW) Level 2 UEFA C and Level 3 UEFA B coaching licence. Importantly, the course attempts to create links between academic and sport-specific coaching knowledge, coaching practice, and experiential learning through a stringent work-based learning design. It is thought that such an environment allows students to learn through interaction with their peers, coaches and mentors, in a community of practice where students can develop their knowledge and expertise by interacting on an on-going basis (Wenger *et al.*, 2002).

2. CURRICULUM DESIGN

The designers of the degree recognised coaching "as a complex social encounter" and it is therefore essential that the design of any curriculum is significantly influenced by the opportunity for students to interact with peers, mentors and significant others. Typically NGB coach education has portrayed coaching as a "knowable sequence" (Usher, 1998) and coaches as "merely technicians involved in the transfer of knowledge" (MacDonald and Tinning, 1995) where theory has been delivered separate to practice, presenting high level tasks as sequential routine, which has resulted in the de-skilling of the coach in terms of the cognitive and human interactions (Jones, 2000; Potrac *et al.*, 2000). In an effort to move away from this traditional "instructional paradigm" typical of NGB coach education, the FnD has been developed to reflect a constructivist approach to learning, whereby students are given the opportunity through work-based learning and coaching to construct knowledge and meaning from their own experiences. This would support Abrahams and Collins (1998) who suggested that coach education should "explicitly challenge candidates to rationalise, and critically reflect on, the particular blend of coaching tools that have been used".

The constructive alignment of the curriculum within the FnD is designed to facilitate learning in that activities within the curriculum are designed for that purpose and should improve the quality of learning and graduates within that profession (Biggs, 2003). The construction of the curriculum recognises that coach education should not be seen as a form of disjointed, ad-hoc learning pathways typical of many coach education programmes, but provide opportunity to develop coaching communities of practice to interact, reflect and develop appropriate personal philosophies of coaching instruction. The embedded coaching qualification modules provide students with up to three hours of contact time over two twelve-week semesters. On the Level Two licence students acquire 72 hours of teaching and on the Level Three 144 hours, clearly an advantage over the ad-hoc approach typical of many coach education awards. The course design recognises that whilst NGB awards may facilitate the development of pre-requisites appropriate to employment, this development alone does not guarantee it. Therefore it is inappropriate to assume that practicing coaches are highly employable on the basis of passing coach education awards alone. Employability derives from the ways in which the student learns from his or her experiences and this is the underlying theme of the curriculum design. Whilst students are able to acquire the key FAW Level 2 UEFA C licence and Level 3 UEFA B licence within key modules, the course design recognises that modern day coaching is multidisciplinary and requires an understanding of many disciplines. The curriculum design recognises that not all students will enter the football profession, and many will pursue alternative careers in sports development, teaching or community coaching. Therefore the curriculum allows students to acquire traditional academic content (i.e. professional knowledge) through key modules such as Sport Development, Coaching and Sport Psychology and underpinning this is a more functional knowledge essential for coaching practice. The modular programme and levels demonstrate the modules offered at each level and the graduation and progression routes of students (Table 1).

3. WORK-BASED LEARNING

A significant number of credits (the equivalent of 360 hours of contact time) is attached to the work-based learning (WBL) module which is conducted either in a professional club or local authority environment. The significant amount of time attached to the WBL reflects the concerns of Bowers (2006) who suggested for a curriculum to be effective, it is beneficial if designed and developed in a way that encourages students to engage in issues that develop depth in intellectual knowledge and the practical skills to solve problems in the real world. The WBL recognises that present day coach education fails to base awards in the complexity of coach interaction and complex circumstances (Jones *et al.*, 2004) and that coaches must be able to link theory to practice. The experiential learning developed by students through the WBL provides opportunities for novice coaches to experience the complexities of coaching in its many forms. Evidence would suggest that both the experience of the coach and encounters with experienced coaches are fundamental to the shaping of coach habitus and coaching practice (Cushion *et al.*, 2003).

Table 1. Modular design of curriculum.

Level 4	Coaching I	Introduction to Sports Development	Introduction to Sports Psychology		Football Coaching I UEFA C Licence	Strength and Conditioning 1	Exercise Physiology
		Students choose either of these modules					
Level 5	Coaching II	Community Sports Development	Advanced Sports Psychology	Football Coaching II UEFA B Licence Practical Delivery	Football Coaching II UEFA B Licence Taught	Work Based Learning 360 hours (2 days a week) coaching at a professional club or working within a Sports Development department.	
	Graduate with Foundation Degree Football Coaching and Performance or progress on to "top-up" year						
Level 6	BSc Sports Coaching and Performance						

Through the work-based learning students develop "Professional Entity" and make connections between work integrated learning and work based environments (Reid and Petocz, 2003). The work-based learning environment provides opportunity for students to construct meaning from their experiences. Jones *et al.* (2004) suggested that inherent in experiential learning is the process of learning how to coach through socialisation within a subculture, where coaches through interaction with other coaches are able to develop a set of coaching values of "how things should be done" (Lyle, 1999). The WBL environment places students under the mentorship of UEFA Level 3 coaches who interact routinely with the students and provide feedback on coaching practice. Additionally weekly tutorial sessions provide the opportunity for focus group discussion and the chance for students to discuss their experiences. This innovative approach to coach education would suggest the experiential learning opportunities offered through this approach are not as Jones and Wallace (2005) suggest "removed from reality", but provide coaches with the opportunity to facilitate the integration of new knowledge into coaching practice (Nelson and Cushion, 2006).

4. REFLECTIVE PRACTICE

Further, the course and in particular the WBL module is underpinned by a culture of reflective practice, a process that has gained an increasing amount of support within sports coaching (e.g. Knowles *et al.*, 2001; Nelson and Cushion, 2006) and related fields (e.g., sport psychology, Anderson *et al.*, 2004; Cropley *et al.*, 2007). Essentially, reflective practice is thought to afford sports coaches the opportunity to explore issues concerning the effectiveness of their practice and thus create a link between professional knowledge and practice (Schön, 1983). Thus, reflection is widely acknowledged as a vital process within coach education.

In attempts to address the utility of such an approach to coach education the students at Glamorgan University on the FnD in Football Coaching and Performance programme ($n = 16$) engaged in a longitudinal intervention-based study, which aimed to: a) better understand the way in which reflective skills can be developed, b) investigate the influence of engaging in structured reflection on coaching practice and professional development, and c) explore the value of reflective communities in helping participants to extract knowledge and learning from their work-based experiences. Participants were initially asked to reflect in a structured manner (Cropley *et al.*, 2007) on coaching experiences ($n = 8$) conducted in partial fulfilment of the FAW Level 2 UEFA C licence. These reflections were then scored using a modified recognition system (Knowles *et al.*, 2001) and their content analysed. Subsequently, a reflective practice training programme, consisting of: workshops ($n = 2$), tutorials ($n = 4$), and reflective peer focus groups ($n = 4$), was administered to the participants during their engagement in the FAW Level 3 UEFA B coaching licence. Participants again engaged in structured reflection following the training programme in a further six coaching sessions. Results indicate that the level (descriptive to critical, Anderson *et al.*, 2004) at which participants were able to reflect significantly increased as a result of the intervention training programme. Inductive and deductive content analysis procedures conducted on the participants' reflective diaries highlighted that such changes in the reflective level directly influenced the content of the reflections. Specifically, participants tended to focus more on their own coaching behaviours and management of the coaching process as opposed to external organisational factors associated with the content and structure of their coaching sessions. Social validation interviews generally supported these findings and further support was provided from the participants for the integration of academic and vocational development underpinned by experiential learning and a reflective culture. Such findings support the need for coach education programmes to embrace reflective practice as a tool for both personal and professional development. Indeed, participants in this study reported that by engaging in reflective practice they were able to make sense of their experiences and address issues concerning the effectiveness of their practice.

Importantly, the success of the course has primarily been supported through the results of this empirical research and the fact that all students ($n = 16$) now hold the FAW Level 2 UEFA C coaching licence (assessments for the Level 3 award are yet to take place). Such an approach to coach education that includes the integration of NGB awards within the modular programme, supported by high quality academic provision ensures that a new breed of coaches is being developed. Indeed, the fragmented, sequential, ad-hoc approach typical of coach education is being re-conceptualised through a more structured, contextualised and supportive environment. Further, the course echoes the views of Cushion *et al.* (2003) of the need to situate the trainees' learning in the practical experience of coaching in an appropriate supportive context, through work-based learning in a variety of contexts.

References

Abrahams. A. and Collins, D., 1998, Examining and extending research in coach development. *Quest*, **50**, pp. 59-79.

Anderson, A. G., Knowles, Z. and Gilbourne, D., 2004, Reflective practice for applied sport psychologists: A review of concepts, models, practical implications and thoughts on dissemination. *The Sport Psychologist*, **18**, pp. 188-203.

Biggs, J., 2003, *Aligning Teaching and Assessments to Curriculum Objectives* (Imaginative curriculum project, LTSN Generic Centre).

Bowers, H., 2006, Curriculum design in vocational education. Abstract presented at *Australia Association for Research in Education – 2006*, Conference 26 to 30 Nov 2006, Adelaide.

Cropley, B., Miles, A., Hanton, S. and Anderson, A., 2007, Improving the delivery of applied sport psychology support through reflective practice. *The Sport Psychologist*, **21**, pp. 475-494.

Cushion, C., Armour, K. M. and Jones, R. L., 2003, Coach education and continuing professional development: Experience and learning to coach. *Quest*, **5**, pp. 215-230.

Jones, R. L., 2000, Towards a sociology of coaching. In *Sociology of Sport: Theory and Practice*, edited by Jones, R.L. and Armour, K. (London: Addison Wesley Longman), pp. 33-43.

Jones, R. L. and Wallace, M., 2005, Another bad day at the training ground: Coping with ambiguity in the coaching context. *Sport Education and Society*, **10**, pp. 119-134.

Jones, R.L., Armour, K.. and Potrac, P., 2004, *Sports Coaching Cultures: From Practice To Theory* (London: Routledge).

Knowles, Z., Gilbourne, D., Borrie, A. and Nevill, A., 2001, Developing the reflective sports coach: A study exploring the processes of reflective practice within a higher education coaching programme. *Reflective Practice*, **2**, pp. 185-207.

Lyle, J., 1999, The coaching process: Principles and practice. In *The Coaching Process: Principles and Practice for Sport*, edited by Cross, N. and Lyle, J. (Oxford: Butterworth-Heinemann), pp. 1-24.

Lyle, J., 2002, *Sports Coaching Concepts: A Framework for Coaching Behaviour* (London: Routledge).

MacDonald, D. and Tinning, R., 1995, Physical education teacher education and the trend to proletarianisation: A case study. *Journal of Teaching in Physical Education,* **15**, pp. 98-118.

Nelson, L. and Cushion, C., 2006, Reflection in coach education: The case of the National Governing Body Coaching Certificate. *The Sport Psychologist,* **20**, pp. 174-183.

Potrac, P., Brewer, C., Jones, R., Armour, K. and Hoff, J., 2000, Toward an holistic understanding of the coaching process. *Quest,* **52**, pp. 186-199.

Reid, A. and Petocz, P., 2003, The professional entity: rebuilding the relationship between students' conceptions of learning and the future profession. Paper presented at the 11th Improving Student Learning Symposium, Hinckley, Leicestershire, 1-3rd September, 2003.

Schön, D. A., 1983, *The Reflective Practitioner* (New York: Basic Books).

Skills Active, 2006, *Assessment of Current Provision: Sport and Recreation* (London: SkillsActive).

Sports Coach UK, 2008, The UK Coaching Framework. www. sportscoachuk.org

Usher, R., 1998, The story of the self: Education, experience and autobiography, In *Biography and Education: A Reader,* edited by Erben, M. (London, Falmer Press), pp. 18-31.

Wenger, E., McDermott, R. and Snyder, W. M., 2002, *Cultivating Communities of Practice* (Boston, MA: Harvard Business School Press).

CHAPTER TWENTY-SIX

The coaching process in elite youth soccer: The players' experiences

Christopher J. Cushion

School of Sport and Exercise Sciences, Loughborough University, Loughborough, UK

1. INTRODUCTION

Coaching is a social process, comprising a series of negotiated outcomes between structurally-influenced agents within an ever changing environment (Cushion and Jones, 2006). As a result, the coaching process is considered the result of dynamic interaction between coaches, athletes and the socio-cultural context. Although the coaching process is constructed of a series of interpersonal relationships (Cushion et al., 2006), the coaching context and process cannot be understood as simply the aggregate of individual behaviour. Players serve a dual role in the coaching process both as an essential part of it but also in its construction; a player's disposition or habitus in an Academy setting exists through practice and because of practice. The nature and construction of players' interactions with each other, the coaches and the club environment are essential in understanding the coaching process. This report therefore focuses specifically on players, their experiences, behaviour and objectives, and how these interacted with the coach and the club environment to form part of the coaching process in a youth Academy context.

The coaching process affects players, not only in terms of skill development, but also in their experience of participation, the behaviours they engage in, the values and attitudes they learn, and the goal priorities that are set. Therefore, an understanding of players' experiences provides insight into the nature of the coaching process and into the relationship between player, coach and club environment. The purpose of this research was to understand something of the dialectic nature of relationships within the coaching process within a professional youth Academy; the coaches and the club shape the players' experiences, whilst the players' experiences influence the coach and the club and ultimately the coaching process. This analysis enables the examination of social actors' behaviour, of social structures and how these relationships are played out within the social arena of professional soccer. The theoretical framework used comes from Bourdieu (1984), particularly his attempts to address the issue of agency and structure in terms of articulating the relations of production between individuals (player, coach), their body and social structure (the clubs and their culture).

2. METHODS

Following approval from the University's ethics committees, thirty-four players and ten coaches participated in a two-season long (August to May, 20 months) ethnography conducted at two professional soccer clubs in the English Premier League. The players were from the two senior youth teams at the club (under 18s and under 16s), and were a combination of full-time players (30 players) and contracted schoolboy players (4 players) looking to secure full-time status. Five coaches participated and were employed by the clubs. All coaches were UEFA qualified and had a minimum of five years' experience coaching at this level.

An ethnography was used to collect the data that included participant observation and interviews. This approach provided an insight into the players' experience at the club. Observations were conducted over periods ranging from 2 to 4 days per week during the seasons in question, and varied in length between 2 hours to all day depending on the schedule of games, training and education. At the end of the second season, a series of in-depth interviews were conducted with the five coaches. In addition, two group interviews per club (4 players per group) were carried out with a random sample of the Academy players.

Data analysis involved three overlapping levels. First, the data from the interviews and field notes were organised following the general principles of grounded theory. This built a system of themes from the unstructured data representing the coaching process within an active, unfolding coaching context. These themes were conceptually grounded both in the ideas and objectives informing the research and in the empirical observations and included the club context, the coach and the players. This report focuses specifically on the theme relating to the players which had the categories: body subjugation and discipline; collectivity, curtailment and pressure; and transition and identity construction. A level of analysis was subsequently employed to situate data within a theoretical framework (Bourdieu and Wacquant, 1996) that enabled a move from concrete description to abstraction. Importantly, this was not a prejudgment about how to read the data but a process of supporting analysis and interpretation.

3. RESULTS and DISCUSSION

3.1 Body subjugation and discipline

The interpretation of the relationship between player, coach and club has strong parallels with those of both Bourdieu and Foucault regarding body subjugation and discipline (Guilianotti, 1999). The club moved the players from routine social relations and relocated them in a somewhat confined social space. The body of the player was subjected to a familiar but more intense and rigid discipline of training and games and was subject to examination by experts for example, coaches (Foucault, 1977; Guilianotti, 1999). In this sense the body, in professional youth soccer, is a bearer of symbolic value.

Bodily or physical capital was central to the player's relationship with the club, the coaches and his own soccer identity. There was a relationship between the development of the body and social location, with the body being central to the acquisition of status and distinction. In the clubs in this case, the 'first year professionals' and those on trial to get a contract, 'trialists', have undeveloped bodies relative to the field, and therefore possess limited amounts of capital, and suffer consequences in terms of status. For the players, the production of physical capital was about developing the body in ways that are recognised as having value in the social field, revolving around playing ability and physical fitness. This physical capital was the gateway to other important forms of social, symbolic and cultural capital.

The players bear the indisputable imprint of their social class (Bourdieu, 1984; Shilling, 1997), in that the players' habitus is formed in the context of their social position at the club and it inculcates them into a world-view, based on, and reconciled to these positions (Shilling, 1997). However, as Giddens (1984) suggested, the individual player is not completely helpless and at the mercy of social forces in this process. The players want to become professionals and want to improve as these players suggest:

> M: 'Just wanna get into the first team, get a better contract'
> S: 'Become a pro, play well in the reserves, score goals, do well in the reserves'
> T: 'To become a better player each day, to become a professional'

By pursuing their own goals, the players engaged in social practices in the coaching process that contribute to the maintenance of existing culture and help to reproduce it. In so doing, they reproduce existing relationships of power and inequality in a struggle for capital, during which cultural meaning is contested, yet works to favour dominant interests (Light, 1999). Those in power, the coaches, control the players, who must be docile and submissive, and are in fact complicit. This complicity is also an essential element of symbolic violence that can only be exerted on a person predisposed through the habitus to accept such behaviour.

3.2 Collectivity, curtailment of individuality/autonomy, pressure

On a day-to-day basis, players were given little autonomy and were treated by the staff like members of an undifferentiated group rather than individuals. The restriction of individuality was evident throughout and wherever players moved their team-mates usually moved with them. Alongside a curtailment of individuality came a lack of privacy; changing, showering and eating were communal experiences.

The players had no input to, or choice about, their daily routine. The coaches decided what training they would do, and how long that training would last. During the season, team-selection and game tactics too, were entirely in the hands of the coaching staff. In addition, during training, and before and after games, the coaches asked for little direct input from the players. In essence, the control exercised by the coaches and the club resulted in the players being denied all decision-making about their professional and occupational experience and, whilst within the confines of the club, their social experience - a situation recognised by the players.

Players familiar with the routines of playing and training from their previous experience did not seem well equipped for the tense and pressurised atmosphere surrounding being an Academy player; and expressed surprise and concern at the highly pressurised atmosphere and its consequences.

'God I look back and I think I've got worse since I've been here. When I was playing Sunday League, no pressure, doesn't matter if you make a mistake. I was playing good football, I think how the pressure has got to me. Thought I'd get used to it but no. Something that perhaps you don't expect is being put under so much pressure.'

The players were under pressure from the constant scrutiny of the coaching staff. Their skills were clearly visible to others and subject to constant judgement by the coaches and fellow players, invoking a constant pressure to perform which, in turn, contributed to a fear of failure that stifled risk-taking and creativity.

3.3 Transition and Identity Construction

Part of the experience in the players' transitions from trainee to professional was enduring the drawbacks of hierarchical insignificance. It was this socialisation experience that contributed to the formation of the player sub-culture that, in turn, influenced players' individual experiences and impacted the coaching process.

Key to this strong and influential player sub-culture was dramaturgy, or impression management (Goffman, 1961). Using impression management, the majority of Academy players presented themselves as submissive and compliant workers while at the same time partaking in both physical and verbal forms of peer group resistance. A central focus for peer group resistance was the concept of being "busy". For the players, not being busy meant conserving effort while training, essentially impression management. Any player who 'acts too eager' or 'does too much like' was labelled as 'busy'. Like instances of 'output restriction' observed in the wider industrial sphere (Collinson, 1992; Parker, 1996) arguably, these means of player regulation were a means by which collective resistance could be expressed towards their lack of control within the coaching process.

The fear of being labelled 'busy' was an instrument of sub-cultural control, stopping players from volunteering and asking questions. For others it extended beyond peer group resistance and into the realms of the ongoing struggle to strengthen their position; 'I think it's just an excuse for people to rip into you'.

Playing ability, and the expression of that ability in games and training, had a property-like nature amongst the players. It was a form of symbolic capital. It imposed a perception upon the players, that when they recognised and observed symbolic capital (i.e., playing ability) it became transformed to cultural or even economic capital within the field (Cicourel, 1995). The players' positions in the field, and their subsequent social trajectory, as determined in part by their acquisition of capital, were ephemeral and can be conceptualised as a transition from Academy player to professional.

The Academy players' transition from first years to professional status involved separation from their previous social condition, a marginal or transitional phase, and a time when the player is incorporated into the new status, that of being a professional player. In a process of transition, the players through their interaction with club and coaches are in a liminal state (Turner, 1967; 1969) or a marginal state of cultural identity. The players are excluded from the fullness of the professional player's lifestyle and have an inferior and often ambiguous role. The players are considered to be between symbolic statuses: not amateurs; not professionals, they are in transit across the symbolic boundaries of the two.

4. CONCLUSIONS

The findings of this research case suggest that while the coach and the club shaped the players' experiences, these experiences influenced the process, forming dialectic relationships throughout. The findings suggest that for the players in this case their experiences of the Academy environment were interconnected and interrelated to and through the coaching process; occurring while they were simultaneously living out their identity, transition and culture.

References

Bourdieu, P., 1984, *Distinction: A Social Critique of the Judgement of Taste* (Cambridge, Mass: Harvard University Press).

Bourdieu, P. and Wacquant, L.J.D., 1996, The purpose of reflexive sociology (The Chicago Workshop). In *An Invitation to Reflexive Sociology*, edited by Bourdieu, P. and Wacquant, L.J.D. (Chicago: University of Chicago Press), pp. 61-215.

Cicourel, A. V., 1995, Aspects of structural and processual theories of knowledge. In *Bourdieu: Critical Perspectives*, edited by Calhoun, C., LiPuma, E. and Postone, M. (Oxford: Blackwell), pp. 89-115.

Collinson, D. L., 1992, *Managing the Shop Floor*, (London: DeGruyter).

Cushion, C. J. and Jones, R. L., 2006, Power, discourse and symbolic violence: The case of Albion Football Club. *Sociology of Sport Journal*, **23**, pp. 142-161.

Cushion, C. J., Armour, K, M. and Jones, R. L., 2006, Locating the coaching process in practice: models '*for*' and '*of*' coaching. *Physical Education and Sport Pedagogy*, **11**, pp. 1-17.

Foucault, M., 1977, *Discipline and Punishment* (London: Peregrine).

Giddens, A., 1984, *The Constitution of Society. Outline of the Theory of Structuration* (Cambridge: Polity Press).

Goffman, E., 1961, *Asylums* (New York: Doubleday).

Guilianotti, R., 1999, *Football; A Sociology of the Global Game* (Cambridge: Polity Press).

Light, R. L., 1999, *Social Dimensions of Rugby in Japanese and Australian Schools.* Unpublished Doctoral Thesis, University of Queensland, Australia.

Parker, A., 1996, *Chasing the Big-Time: Football Apprenticeship in the 1990's.* Unpublished Doctoral Thesis, Warwick University, UK.

Shilling, C., 1997, *The Body and Social Theory*, (London: Sage).

Turner, V. W., 1967, *The Forest of Symbols* (New York: Cornell University Press).

Turner, V. W., 1969, *The Ritual Process* (London: Routledge and Kegan Paul).

CHAPTER TWENTY-SEVEN

Understanding the coaching process in elite youth soccer

Christopher J. Cushion

School of Sport and Exercise Sciences, Loughborough University, Loughborough, UK

1. INTRODUCTION

Coaching is a social activity built on a web of complex context- and inter-dependent activities that come together to form a holistic coaching process (Lyle, 2002; Jones *et al.*, 2004; Cushion *et al.*, 2006). Although the coaching process has been recognised within the literature for over two decades, an examination of existing research indicates that an in-depth understanding remains lacking (Cushion *et al.*, 2006; Cushion, 2007). There is evidence to suggest that coaches work without reference to any process "model", preferring instead to base their coaching on experience, feelings, intuitions and events to trigger actions (Lyle, 2002; Cushion *et al.*, 2006; Jones *et al.*, 2004). However, Lyle (2002) argued that "improvements to coach education and to coaching depend on a sound understanding of the coaching process" (p. 29). Further clarification of this multifaceted process would appear to be a very necessary step before establishing realistic guidelines for good practice (Lyle, 2002; Cushion *et al.*, 2006).

Unfortunately, analysis of the coaching process is under-developed, with relatively few empirical studies having emerged. Instead we have a broad range of research that, while useful, remains limited because it fails to grasp the dynamic, complex, messy reality of coaching (Cushion *et al.*, 2006). Instead, coaching is portrayed as a series of asocial, linear and unproblematic episodes. This narrow, reductionist, rationalistic and bio-scientific approach to coaching not only reduces the complexity of practice, but assumes incorrectly that separating and truncating components of real life coaching, then feeding them back to coaches to make sense of, can result in more than abstractions that become a substitute for real life (Cushion, 2007). Such an approach can never fully account for what coaching actually entails.

Moreover, because coaching can be readily represented as "episodes" and therefore parts of it described in individual terms, it is easy to overlook the degree in which the inter-relatedness and inter-connectedness of coaching sustains the process (Jones *et al.*, 2004; Cushion *et al.*, 2006; Cushion 2007). Conceiving a social structure such as coaching as the "mere aggregate of individual strategies and acts of classification makes it impossible to account for their resilience as well as for the emergent objective configurations these strategies perpetuate or challenge" (Bourdieu and Wacquant, 1992, pp. 9-10). This relational nature makes the actuality of coaching and its process significantly different to representations of it. Typical representations lead us to reduce the effect of the context and reduce coaching to the outcome of direct action actualised during an interaction (Cushion, 2007). As such, we understand much about the behavioural 'what' of coaching but less about the 'why' and 'how'.

While Wenger (1998) argued that simple practitioner models offer succinctness and portability, he also suggested that such approaches can ossify practice around their inertness, thus hindering the very conceptual and practical development they are designed to promote (Cushion, 2007). Moreover, coaching models developed from an immature or limited understanding can hide meaning in blind sequences of operation, with the knowledge of a formula or schema leading to the illusion that one fully understands the process that it describes (Wenger, 1998; Cushion, 2007). To the practitioner though, these representations of the coaching process frequently seem disconnected from their context and frozen into text that fails to capture the reality of the coaching world. Consequently, overly simplistic representations of coaching may even be viewed with cynicism by practitioners and scholars alike, a somewhat ironic substitute for that which it is intended to reflect (Wenger, 1998; Cushion, 2007), and as such are discarded or not utilised at all.

The purpose of this research was to address some of these issues and engage in understanding, documenting and analysing the complexity inherent within the coaching process in elite youth soccer. The intention being to better inform both coaching and coach education. A more holistic view is employed by engaging in a research programme embedded *in* coaching, and to think with greater depth and detail to increase an understanding *of* coaching.

2. METHOD

Following ethical approval, from the University's ethics committees, thirty-four players and ten coaches participated in a two-season long (August to May, 20 months) ethnography conducted at two professional soccer clubs. The players were from the two senior youth teams at the club (under 18s and under 16s), and were a combination of full-time players (30 players) and contracted schoolboy players (4 players) looking to secure full-time status. Five coaches participated and were employed by the clubs, were UEFA qualified and had a minimum of five years' experience coaching at this level. In addition, semi-structured interviews were conducted with five 'expert' youth soccer coaches to triangulate the data from the case study clubs and add another interpretive layer to the findings. Both clubs were English Premiership football clubs (the highest professional football division in England).

Data were collected within an ethnographic framework that included participant observation and interviewing. This approach enabled an insight into the varying and evolving coaching process at the club. Observations were conducted over periods ranging from 2 to 4 days per week during the seasons in question, and varied in length between 2 hours to all day depending on the schedule of games, training and education. At the end of the second season, a series of in-depth interviews were conducted with the five coaches. In addition, two group interviews per club (4 players per group) were carried out with a random sample of the academy players.

The data analysis involved an iterative shift from data collections to analysis and description and to write-up and theory (Cushion and Jones, 2006). The process involved three overlapping levels. First, the data from the interviews and field notes were organised following the general principles of grounded theory. This built a system of themes from the unstructured data representing the coaching process within an active, unfolding coaching context. These themes were conceptually grounded both in the ideas and objectives informing the research and in the empirical observations and were the club context, the coach and the players. Second, these themes were used to produce a descriptive account of the coaching process. Although these descriptions highlighted the various relationships under study, they did not capture the true complexity of coaching at the clubs (Cushion and Jones, 2006). Consequently, a third level of analysis was employed to situate data within a theoretical framework (Bourdieu and Wacquant, 1996) that enables a move from concrete description to abstraction. Doing so increased understanding of the range and variability of the social actors (coaches and players) and structures (the professional club environment) under study and how they interacted to create a problematic coaching process (Cushion and Jones, 2006). Importantly, this was not a prejudgment about how to read the data but a process of supporting analysis and interpretation.

3. RESULTS and DISCUSSION

- *The coaching process is not necessarily cyclical, but is continuous and interdependent and operates with objectives that exert a continuous influence. These objectives derive from the club, the coach and the players.*

The research in this case challenges the notion of coaching being a linear or cyclical episodic process. The data demonstrated a complex web of clustered context-dependent and interdependent relationships and actions. The workings of the process in terms of this interdependence and interconnectedness can best be likened to dropping a pebble into a pond, with ripples moving out and back throughout. Hence, action in one area would impact the whole but not always in a direct and linear fashion, instead the impact would often be indirect and operate in a dialectic fashion. Therefore, clustered around the coach, the player, and the club environment, coaching practice is socially constructed and deeply embedded in social and cultural contexts. The coaching process represents a constructed *relationship between* the coach, the player, and the club environment. The crucial part to understanding the coaching process is the *relationship between,* as neither, coach, player, or club has the capacity to determine action unilaterally. It takes the meeting of disposition and position, the correspondence between mental structures and social structures, to generate practice (Bourdieu and Wacquant, 1996).

3.1 The club environment

The Academy structure (itself positioned hierarchically within the club) was rigidly hierarchical with clear differentiation between the coaches and players, the coaches themselves and between, first, second and third year players. In addition, there was a strong organisational culture focused on the needs of the club's first team, that in addition to the hierarchy impacted directly and indirectly upon the working practices of both coaches and players, as this quote illustrates: "The first team do what they do and get what they want and everything is affected by them."

For the Academy and the club, winning and the importance of winning, were held in high esteem and were remarkably influential parts of the culture as this coach suggests: "...you can get more if you win." Moreover, while the espoused philosophy was deemed as "player centered", the primary concern was a steady stream of players continuing to emerge from the Academy to support and supplement the needs of the first team and reserves. The approach was clearly club-centered, not player-centered as this coach confirms: "The objective for the club and first team manager is to get players for his team."

The club environment therefore, through its hierarchical structure, its objectives and culture is both part of and creator of the coaching process. The club serves as a social force actively shaping the coaching process, what is coached and why it is thought that these things are important.

3.2 The coach

The coach's experience was highly influential in shaping the nature of day-to-day practice within the coaching process. The body acting as a social memory, "what coaches do", signifies much about their personal history. Knowledge and action are the product and manifestation of a personally experienced involvement in the sport. In this study being an ex-professional player influenced coaching practice as this coach suggests: "some of the things I have now, I thought, I do this because I did it then as a player". The cultural capital associated with being a former player contributed strongly to a cultural hierarchy that often outweighed or over-ruled the more formal social hierarchy.

The perception of the coach's role was also influential. The desire to win was often fuelled by coaches' need and desire to build and enhance their reputation, or their cultural capital. The objectives of practice therefore, were not linked to player development but acted as a stepping stone to first team coaching, as is suggested by this excerpt of data: "I will move on, hopefully I will move on to bigger and better things."

The coach is at the heart of the coaching process, occupying a position of centrality and influence. The coaches are shaped by the coaching process through their previous experience and their interactions with the players and the club, while acting as a shaper of the process often with the intention to improve their position within football.

3.3 The players

The players were not simply empty and passive vessels receiving coaching dogma. They, not unlike the coach and the club, were both shaped by and shapers of the coaching process. The players had to structure their behaviour around coach and club defined explicit and implicit dictates, while at the same time acting back by seeking to improve their position within the culture, often at the expense of one another, as this player observes; "It's dog eat dog really, you've got to look after yourself". More, the players maintained a strong player subculture, utilising peer group resistance to impact the functional coaching programme, what was actually taught to the players: "You've got to get the players ready to work hard first, before you even think about what you'll do in that session". The player subculture and the players themselves were powerful transmitters of differential treatment and values. This resulted in a hierarchy within a hierarchy determined by playing ability, a form of physical and symbolic capital. The players contributed dialectically to power relations and culture and ultimately to the delivery and effectiveness of the coaching process.

- *The process is a constantly dynamic set of intra- and inter-group interpersonal relationships. These relationships are locally dialectic between and amongst agent (coach, player) and structure (club, culture) and are subject to a wide range of pressures. These pressures derive from the coach, the players and club and are embedded in the culture and traditions of the sport.*

The coaching process is underpinned by complex interaction. The interactions identified within this research are not benign and are illustrative of a rhetorical ideal of togetherness of the players and common purpose between coaches and players. There was the underlying nature of antagonistic groups struggling to impose a definition of the world most congruent with their interests (Bourdieu and Wacquant, 1996). Thus, the coaching process and experiences within it were uneven depending on the player's formal position within the year group structures, and informally against coach partiality and selectivity, which meant in practice sub-groups of players existed. Membership of a particular group meant a differing relationship with the coach and a different experience from, and contribution to, the coaching process as the players describe in the excerpt of data:

T: "The way some people are spoken to is different"
M: "Off the pitch as well, just walking round the place. Andy [pseudonym] will speak to some people more, Greg [pseudonym] will speak to some people more"
A: "I've noticed that Andy will say alright to everyone, but to some player it will be a bit more, it will be alright, and how's the leg, and will you be back for Saturday"
T: "In games and training too, player A makes a mistake no problem, player B

makes a mistake, all over them"

S: "Yeah"

T: "Yeah N makes a lot of mistakes"

M: "Yeah"

A: "But a mistake someone else makes will get ripped, but he does it and it's alright"

- *The coaching process is embedded within external constraints, only some of which are controllable.*

The influence of the club has already been touched on, but the Academy existed within a broader club environment that imposed external constraints on the operation of the coaching process for the Academy coaches. These constraints could be around the use of facilities in terms of access, or more commonly as a function of the hierarchy. Both contexts were affected by an ongoing campaign by the first team manager to reduce the number of professionals, especially those who were not in his future plans. These groups, often with only five or six players, would have to train on their own. The reserve team manager, using his seniority, could apply pressure on the Academy coaches, which meant that the reserves would join in with them as these data illustrate:

Academy Manager: "When was the last time we did some coaching, this game is catering to the needs of the reserve team. It seems to me that since Christmas we're not really doing any structured coaching sessions. I get ambushed by reserve team coach who says do you want a game, I don't really but he is the reserve team manager reporting into the first team manager so I'm caught having to agree with him. I think because he only has 5 players he gets caught out a little about thinking what to do with them. The manager wants them gone as soon as possible so it must be hard for him, trying to keep the manager happy."

- *A pervasive cultural dimension infuses the coaching process through the coach, the club, the players and their interaction.*

The coaching process in this case was "ideologically laden" (Jarvie and Maguire, 1994, p. 184). The coaches were not simply passing on the products of soccer (technical/tactical), but were instilling attitudes, values and beliefs and giving the players experiences suited to the professional game, thus reproducing the culture. The Academy, the coaching context, through its symbolic strength has the capacity to inculcate learning successfully as a function of its weight in the structure of power relations. The coaching process was used to impose the language, meanings, symbolic systems and its culture. The players' desire to succeed in the field ensures that this activity is misrecognised, and the arbitrary but legitimate culture is experienced as an axiom; players and coaches no longer question 'Why?'

4. CONCLUSION

We sought to further understanding of the coaching process. We attempted to grasp the complexity of the coach, the player and the club environment, investigating the dynamic relationship between these factors and the implications of this interaction for coaching practice and the coaching process. The findings from this study and others that have contextualised analyses of practice suggest a need to question existing conceptions of the coaching process and develop evidence for others (e.g., Saury and Durand, 1998). Paying attention to the detail of coaching practice, and the social forces described in this study that shape coaching practice and the interconnections that run between them, has revealed much about the construction and complexity inherent in the process (see also, Cushion *et al.*, 2006). In particular, and in conflict with currently recognised models of coaching and coach education, it seems worthwhile reiterating that it is unlikely that coaching can be reduced to the application of generic rules.

As practice unfolded in this case, it seemed to evolve not in a linear or formulaic way but organically, thus tending to operate just beyond formal description and control. As a result, the landscape of coaching practice does not, arguably, align well with structures of division and boundaries. That is not to say that coaching is not independent of structures, but crucially neither is it simply reducible to them. The boundaries of coaching practice in this case were constantly renegotiated between the club, the coaches and the players, defining much more fluid and textured forms of activity that in turn required both a sophisticated and detailed analysis. As Cushion (2007) has contended, without tackling crucial questions about the nature of coaching practice and being immersed in that practice we are likely to develop representations of the coaching process that are systematic distortions of both knowledge and understanding.

References

Bourdieu, P. and Wacquant, L. J. D., 1992, *An Invitation to Reflexive Sociology*, (Chicago: Chicago University Press).

Bourdieu, P. and Wacquant, L. J. D., 1996, The purpose of reflexive sociology (The Chicago Workshop). In *An Invitation to Reflexive Sociology*, edited by Bourdieu, P. and Wacquant, L. J. D. (Chicago: University of Chicago Press), pp. 61-215.

Cushion, C.J., 2007, Modeling the complexity of the coaching process. *International Journal of Sport Science and Coaching*, **2**, pp. 395-401.

Cushion, C. J. and Jones, R. L., 2006, Power, discourse and symbolic violence: The case of Albion Football Club. *Sociology of Sport Journal*, **23**, pp. 142-161.

Cushion, C. J., Armour, K, M. and Jones, R. L., 2006, Locating the coaching process in practice: models '*for*' and '*of*' coaching. *Physical Education and Sport Pedagogy*, **11**, pp. 1-17.

Jarvie, G. and Maguire, J., 1994, *Sport and Leisure in Social Thought*, London: Routledge.

Jones, R.L., Armour, K. and Potrac, P., 2004, *Sports Coaching Cultures* (London: Routledge).

Lyle, J., 2002, *Sports Coaching Concepts: A Framework for Coaches' Behaviour* (London: Routledge).

Saury, J. and Durand, M., 1998, Practical knowledge in expert coaches: On site study of coaching in sailing. *Research Quarterly for Exercise and Sport,* **69,** pp. 254-266.

Wenger, E., 1998, *Communities of Practice: Learning Meaning and Identity* (Cambridge: Cambridge University Press).

CHAPTER TWENTY-EIGHT

An expert English youth soccer coach's beliefs on decision-making and its development in soccer: A case study

Martin I.M. Dean[1] and Christopher J. Cushion[2]

[1]School of Sport and Education, Brunel University, UK
[2]School of Sport and Exercise Sciences, Loughborough University, UK

1. INTRODUCTION

The study of sport expertise is concerned with both identifying characteristics that distinguish experts from the less-skilled and characteristics of the learning environment in which expertise is acquired. From a perceptual-cognitive perspective, decision-making skills become increasingly more important as performers reach higher levels of excellence. This increasing importance is due to higher homogeneity regarding their physical and physiological characteristics at these levels (Reilly *et al.*, 2000; Williams and Reilly, 2000). It is thought that perceptual-cognitive and perceptual-motor streams are in constant interaction between perceiving and understanding and perceiving and doing, with the result being either performance constraint or performance facilitation (Starkes *et al.*, 2004). One example of this could be represented by a young soccer player seeing an effective run made by a team mate but not having the perceptual-motor skill to turn this into a suitable execution fast enough. The counter to this could be that a young soccer player has a well-honed technique and relies on this due to not having the perceptual-cognitive skill to see a more appropriate option and solution. In developing expertise, the literature suggests that the design of practice should consider not only the volume but also the content, as the type that determines potential for accelerating skill acquisition and performance (Ericsson *et al.*, 1993). Further to this, practice which is specifically directed at developing perceptual-cognitive and perceptual motor skills is thought to be most beneficial (Abernethy, 2008).

The aim of this study was to examine the beliefs, values and knowledge of an expert youth soccer coach regarding decision-making and its development. A case study framework was utilised, in an attempt to develop an inductive, 'ground-up' investigation via the construction of expert knowledge (Jones *et al.*, 2003), which when mapped against empirical findings could be used to mediate between the languages of empiricism and craft knowledge. Over the last twenty years, heuristic investigation methods such as ideographic approaches have been more increasingly called for to obtain a better understanding of human behaviours in the sport and exercise domain (Cote *et al.*, 1993). The purpose of grounded theory is to develop concepts and theories based on behaviours of individuals rather than to prove or disprove existing theory (Glaser and Strauss, 1967). Therefore in summary, the overall purpose of this study was to rationalise (Cote *et al.*, 1995) one aspect of the complex, integrative and multi-faceted domain of expertise by identifying key variables that could have a positive impact on decision-making and its development in soccer. Data were collected using an in-depth semi-structured interview to elicit responses in four key areas: player characteristics, coach characteristics, the learning environment and match-play.

2. METHODS

2.1 Participant

The participant in this study was an expert soccer youth coach whose selection for the study was based on the purposive sampling method. The coach was identified as someone who could be expected to offer rich knowledge within the area of interest from a grounded perspective. The coach (age = 44 years) was an ex-professional player ($n = 17$ years), is a UEFA A Licence Coach, a full-time youth coach ($n = 7$ years) with experience at the highest level of youth football in England as a Premier League Youth Academy Coach ($n = 3$ years).

2.2 Interview

A semi-structured interview was used to explore four main areas: (a) Player characteristics, (b) Coach characteristics, (c) The learning environment and (d) Match-play. The interview was transcribed verbatim.

2.3 Data analysis

The objective of the analysis was to identify categories that emerged from the data. This represented a focused method for presenting an expert coach's knowledge regarding the development of decision-making at elite youth level. Data were organised and analysed based on the principles, procedures and techniques of grounded theory (Strauss and Corbin, 1998). A process of open coding was conducted that involved data being broken down, examined and compared for similarities and differences to develop concepts. For example, quotes of interest relating to "Player characteristics" were highlighted and followed by an attempt to distinguish between different types. As analysis continued, quotes were selected and placed in newly developed conceptual sub-categories that represented a step down from the more abstract, higher-order and global categories such as "Player characteristics" that were used within the interview framework.

3. RESULTS and DISCUSSION

Results confirmed the four central aspects to the development of decision-making for elite young soccer players: player characteristics, coach characteristics, the learning environment and match-play. Each aspect is described and explained using sub-categories and associated descriptive concepts.

3.1 Player characteristics

This main category was described by the sub-categories of: Intense Curiosity, Strong Inner-Belief and Acute Awareness and Recognition.

3.1.1 Intense Curiosity

"The best have a real thirst for knowledge and are willing and prepared to ask questions due to the real desperation to drive themselves on."

"The best players and those who badly wanted to learn would say 'why am I doing this? for what reason? individually why? For the collective good – why?'"

"The special people I've worked with are always people who are willing and prepared to ask questions (not in a negative way) of coaches – why are we doing that?, where, when and how, to try and improve themselves."

Winner (1996) argued that only children with the "rage to master" could make it through the gruelling years of training needed to achieve expert ability. The rage to master may be the point at which nature unequivocally makes its constraints felt and sees individuals manipulate their environments in order to render them intellectually stimulating. On the subject of motivation, Ericsson and Lehman (1996) pointed out, experts acquire expertise through extensive "deliberate practice", designed by a coach to improve specific aspects of performance through repetition and successive refinement. One can only assume that the investment required for this would have to be driven by an intense intrinsic motivation.

3.1.2 Strong Inner-Belief

"They have confidence in themselves and a belief."

"The best players – they never lose that focus" ....("he's not gonna score", "he's not gonna get across me wherever the ball is")

According to Young and Medic (2008), the intense and enduring commitment that elite performers in sport are known to show (see Scanlan *et al.*, 1993) could be due to the level of perceived competence a performer possesses and furthermore, to the fact that high task-oriented athletes (see progress in relation to their own previous standards) perform better and spend more time practising than those who have ego goals (outperforming others).

3.1.3 Acute Awareness and Recognition

"The better players can retain information and recognise quicker as youngsters what gets them success."

"With good players it goes in there and it stays."

"Good decision-makers pick the right one more often."

"The best will retain information – it might be a little piece of information but they do retain information and can make the right decision under pressure."

"The best players see those things and think 'yeah it works' and if I do that then I'll get in there."

This response concurs with the theory of long-term working memory (Ericsson and Kintsch, 1995) which accounts for the superior memory that expert athletes show and which extends to response selection, response alteration and response execution. The premise is that processes utilised for eliciting cues can be the same for selecting the appropriate response. Moreover, that retrieval structures which enhance both storage and accessibility enable appropriate prioritised actions and alternative actions to be produced. Pattern recall (Farrow and Raab, 2008), pattern recognition (Williams and Davids, 1995), consistent response selection, together with the ability to determine the likelihood ratio for each option presented (Ward and Williams, 2003) are all aspects that may offer further insight to the findings.

3.2 Coach characteristics

This main category was described by the sub-categories of: Respect and Trust, Encourage Self-Analysis and Deep Thinkers.

3.2.1 Respect and Trust

"Give players freedom with responsibility. I think that incorporates for me everything you're looking for."

"Something he'd shown us [the coach] that we ended up getting some success out of was reinforced by results – this can change people."

"If players have seen a result from doing things in a certain way it just reinforces what you've said. I think the important part to this is that as a result they learn to trust you."

3.2.2 Encourage Self-Analysis

"The best coaches question you so that you start asking questions about yourself and make you start thinking in more depth about the game."

"He was always saying; could you have done that differently? Could you have done that better? What about if you'd have done this? What about if you'd have done that? Always questioning you and making you think about your own game rather than telling you that you should have done this or must do that."

"Suddenly, players are not just thinking about their own game but also thinking about the players that they are playing with. For example, how what you are doing impacts upon them."

3.2.3 Deep Thinkers

"You've got to go into it and really think deeply about how you're going to get the best out of players in that session."

"Coaches must be aware that there are lots more things other than individual strengths that come into decision-making. For example: the quality of the opposition you're playing against, the system you're playing, the strategy you're playing within, the weather….."

Based on the above, one important message for coaches is as follows:-

"learners should be viewed as active problem-solvers rather than "empty vessels" or passive recipients of information" (Williams and Hodges, 2005)

In accordance with the data, the thread here is that deep and careful consideration must be given to the amount and type of feedback offered, albeit within the much wider sphere of determining how best to design, implement and evaluate training programmes that are capable of optimising the decision-making development of the learner.

In summary, research is beginning to show the distinct advantages of having access to an expert coach (Baker *et al.*, 2003), one hallmark of which is meticulous planning of practice. It has been shown that expert coaches spend larger amounts of time planning practices and are more precise in their goals and objectives for sessions than non-expert coaches (Voss *et al.*, 1983).

3.3 Learning environment

This main category was described by the sub-categories of: Balanced Range of Game Aspects, Systems of Play Education, Strong Message Reinforcement and Highly Challenging and Stimulating Games.

3.3.1 Balanced Range of Game Aspects

"It is important to have a coaching syllabus that you believe covers all avenues and will work for all players."

"I think you have to get the right balance between technical, tactical, physical and mental."

"Through the coaching syllabus, it is important to look at every avenue for how to give all players every opportunity to be the best they can be and comfortable in different systems both defensively and in attack."

3.3.2 Systems of Play Education

"If you make the goalkeeper kick it every time he gets it, how are the back four ever going to learn how to play out from the back?"

"At a certain age, whatever session is put on must be able to have an effect on the individual as a part of the collective and within this, guidance on roles and responsibilities becomes ever more important."

3.3.3 Strong Message Reinforcement

"By nature of the structure and the layout of the practice, the coach must know how best to get the desired messages out of a session for the benefit of all of the players."

"As a coach, you'll just keep reinforcing things with enthusiasm. For example, where are you going to get the ball? How are you going to get the ball? What options have you got? Yeah, brilliant - you've played around the corner and followed. What about when you did that? Look at the space it made for you."

3.3.4 Highly Challenging and Stimulating Games

"I like sometimes throwing those in and really making people make decisions and to be really thinking and concentrated on lots of different things because players have to in matches."

"There are days where I like to work the technical, tactical, physical and mental together and will put games on that enable this. You have to multi-task to get them all in."

In relation to the above categories, as Patterson and Lee (2008) suggested it is extremely important to stay mindful of the strong link between information-processing effort given in practice and the level of skill expertise developed in return. In accordance with this, the above data offer the recommendation of a challenging, engaging and problem-solving learning environment that simulates the requirements of match-play.

3.4 Match-Play

The main category was described by the sub-categories of: Effective Learning Strategies in Match-Play and Match-Play as the Ultimate Test.

3.4.1 Effective Learning Strategies in Match-Play

"An effective approach can be to 'command' that players play in a certain way as an experiment for the purpose of later discussion".

"I think you get to a certain age where the individual has to be part of the collective."

The above quotes are in accordance with Abernethy's (2008) philosophy of youth development programmes underpinned by routine task variation and encouragement of a sense of experimentation. In relation to this, presenting a range of decision-making scenarios in line with the future demands of professional soccer may represent an optimal player development opportunity.

3.4.2 Match-Play as the Ultimate Test

"It is during match-play that the best decision-makers come to the fore."

"When you play in matches against different teams, even at academy level you are under pressure. You are under pressure to impress your coach, the academy manager and ultimately you have to be able to respond to that pressure. The best players do and are able to take that pressure and use it as a positive rather than a negative."

Singer and Janelle (1999) addressed the need for competition-specific practice by highlighting the influence of varying levels of performance-related anxiety, anger, joy and other emotions on information-processing and attentional demands. The issue is thought to be not one of ability but more of negotiating problems regardless of situational circumstances that arise in competition. The critical point is that one would expect experts to possess superior coping skills within the typically stressful competition environment.

Not getting away from the match-play environment being the 'ultimate test', the famous basketball coach, John Wooden's belief was that players should be encouraged to focus on the process rather than the end result, to both reduce anxiety and aid performance in pressure situations (Horton and Deakin, 2008). It has been suggested that in pressure situations, if anxiety is present, self-consciousness often follows and in turn, focus is turned inward to form a kind of 'paralysis by analysis' (Jackson and Beilock, 2008). Maybe this is somehow avoided more often by the most consistent and advanced decision-makers (Gallwey, 1974).

4. CONCLUSION

This research represents an informative first step in developing a 'ground-up' perspective of the determinants of developing decision-making in soccer. By using a case study approach, it has been possible to make connections between 'craft knowledge' and certain reported empirical findings.

When considering the construction of a soccer learning environment, the concepts revealed in this study hold the potential to be manipulated by other coaches in their own coaching processes in order to facilitate the development of decision-making in soccer. Researchers are promoting the need for practice and instruction processes to be based on scientific evidence from motor learning and not coach emulation, tradition, or intuition (Williams and Hodges, 2005) and upon consideration of this, this study might be seen to have two main limitations. First, although the coach is an expert, the results are only the beliefs of a single coach and may not represent the general beliefs of the wider expert coach community. Second, although emulation of other coaches provides good examples, it can also provide those that are one-dimensional. However, this particular study does provide rich and meaningful subjective examples of how a vastly experienced practitioner perceives, creates and interprets his world and was designed to relate this to scientific evidence. In future researchers could include further investigation into existing 'craft knowledge' best practice regarding the optimal development of decision-making, for the purpose of identifying what does and does not fit with current theory. Findings from a gap-analysis approach could then be further examined from a variety of perspectives to contribute to more specialist and sophisticated training recommendations for the development of decision-making in soccer.

References

Abernethy, B., 2008, Introduction – Developing expertise in sport – how research can inform practice. In *Developing Sport Expertise: Researchers and Coaches put Theory into Practice*, edited by Farrow, D., Baker, J. and MacMahon, C. (London: Routledge), pp. 1-14.

Baker, J., Horton, S., Robertson-Wilson, J. and Wall, M., 2003, Nurturing Sport Expertise: Factors influencing the development of the elite athlete. *Journal of Sports Science and Medicine*, **2**, pp. 1-9.

Cote., J., Salmela, J. and Baria, A., 1993, Organizing and interpreting unstructured qualitative data. *The Sport Psychologist*, **7**, pp. 127-137.

Cote, J., Salmela, J., Trudel, P., Baria, A. and Russell, S., 1995, The coaching model: A grounded assessment of expert gymnastics coaches' knowledge. *Journal of Sport and Exercise Psychology*, **17**, pp. 1-17.

Ericsson, K.A. and Lehman, A.C., 1996, Expert and exceptional performance: evidence of maximal adaptations of task constraints. *Annual Review of Psychology*, **47**, pp. 273-305.

Ericsson, K.A., Krampe, R. and Tesch-Romer, C., 1993, The role of deliberate practice in the acquisition of expert performance. *Psychological Review*, **100**, pp. 363-406.

Ericsson, K.A. and Kintsch, W. 1995, Long-term working memory. *Psychological Review*, **102**, pp. 211-245.

Farrow, D. and Raab, M., 2008, A recipe for expert decision-making. In *Developing Sport Expertise: Researchers and Coaches put Theory into Practice*, edited by Farrow, D., Baker, J. and MacMahon, C. (London: Routledge), pp. 137-154.

Gallwey, T., 1974, *The Inner Game of Tennis* (London: Pan Books).

Glaser, B.G. and Strauss, A. L., 1967, *The Discovery of Grounded Theory: Strategies for Qualitative Research* (Chicago: Aldine).

Horton, S. and Deakin, J.M., 2008, Expert coaches in action. In *Developing Sport Expertise: Researchers and Coaches Put Theory into Practice*, edited by Farrow, D., Baker, J. and MacMahon, C. (Oxon: Routledge), pp. 75-78.

Jackson, R.C. and Beilock, S.L., 2008, Performance pressure and paralysis by analysis: research and implications. In *Developing Sport Expertise: Researchers and Coaches put Theory into Practice*, edited by Farrow, D., Baker, J. and MacMahon, C. (Oxon: Routledge), pp. 104-118.

Jones, R.L., Armour, K.M. and Potrac, P., 2003, Constructing Expert Knowledge. A case study of a top level professional soccer coach. *Sport Education and Society*, **8**, pp. 213-229.

Patterson, J.T. and Lee, T.D., 2008, Organizing practice: the interaction of repetition and cognitive effort for skilled performance. In *Developing Sport Expertise: Researchers and Coaches put Theory into Practice*, edited by Farrow, D., Baker, J. and MacMahon, C. (Oxon: Routledge), pp. 119-134.

Reilly, T., Williams, A.M., Nevill, A. and Franks, A, 2000, A multidisciplinary approach to talent identification in soccer. *Journal of Sports Sciences*, **18**, pp. 668-676.

Scanlan, T.K., Carpenter, P.J., Schmidt, G.W., Simons, J.P. and Keeler, B., 1993, An introduction to the Sport Commitment Model. *Journal of Sport and Exercise Psychology*, **15**, pp. 1-15.

Singer, C. and Janelle, C., 1999, Determining sport expertise: From genes to supremes. *International Journal of Sport Psychology*, **30**, pp. 117-150.

Starkes, J.L., Cullen, J. D. and MacMahon, C., 2004, A life-span model of the acquisition and retention of expert perceptual-motor performance. In *Skill Acquisition in Sport – Research, Theory and Practice*, edited by Williams, A. M. and Hodges, N.J. (Oxon: Routledge), pp. 259-281.

Strauss, A. and Corbin, J., 1998, *Basics of Qualitative Research: Techniques and Procedures for Developing Grounded Theory*, 2nd edn. (Newbury Park, CA: Sage).

Voss, J., Green, T. and Penner, B., 1983, Problem-solving in social sciences. In *The Psychology of Learning and Motivation: Advances in Research Theory*, edited by Bower, G. (New York: Academic Press), pp. 165-213.

Ward, P. and Williams, A.M., 2003, Perceptual and cognitive skill development in soccer: the multidimensional nature of expert performance. *Journal of Sport and Exercise Psychology*, **25**, pp. 93-111.

Williams, A.M. and Davids, K., 1995, Declarative knowledge in sport: a byproduct of experience or a characteristic of expertise? *Journal of Sport and Exercise Psychology*, **7**, pp. 259-275.

Williams, A.M. and Hodges, N.J., 2005, Practice, instruction and skill acquisition in soccer: Challenging tradition. *Journal of Sports Sciences*, **23**, pp. 637-650.

Williams, A.M. and Reilly, T., 2000, Talent identification and development in soccer. *Journal of Sports Sciences*, **18**, pp. 657-667.

Winner, E., 1996, The rage to master: The decisive role of talent in the visual arts. In *The Road to Excellence: The Acquisition of Expert Performance in the Arts and Sciences, Sports and Games*, edited by Ericsson, K.A. (Mahwah, NJ: Lawrence Erlbaum Associates), pp. 271-301.

Young, B.W. and Medic, N., 2008, The motivation to become an expert athlete – How coaches can promote long-term commitment. In *Developing Sport Expertise: Researchers and Coaches Put Theory into Practice*, edited by Farrow, D., Baker, J. and MacMahon, C. (Oxon: Routledge), pp. 43-59.

Index